CALCULATIONS II
By Tom Henry

Copyright©2002 by Tom Henry. All rights reserved. No part of this publication may be reproduced in any form or by any means: Electronic, mechanical, photocopying, audio or video recording, scanning, or otherwise without prior written permission of the copyright holder.

•This book is *coded* for Copyright© laws of this material. Violators will be prosecuted subject to the penalties of the Federal Copyright Laws.

United States of America
Register of Copyrights
Library of Congress
Washington, D.C.

While every precaution has been taken in the preparation of this material to ensure accuracy and reliability of the information, instructions, and directions, it is in no way to be construed as a guarantee. The author and publisher assumes no responsibility for errors or omissions. Neither is any liability assumed from the use of the information contained herein in case of misinterpretations, human error or typographical errors.

National Electrical Code® and NEC® are Registered Trademarks of the National Fire Protection Association, Inc., Quincy, MA.

ISBN 0 - 945495 - 54 - 4

ENRY PUBLICATIONS SINCE 1985

CALCULATIONS II WORKBOOK

I'm sure you'll find Calculations II a very interesting and factful book to help you prepare for the electrical examination.

I now have written over twenty books to help electricians in their studies. My biggest seller "Calculations For The Electrical Exam" has been a favorite of the electricians for several years. I now have put together, in this book, 101 of the favorite calculation questions asked on electrical exams throughout the U.S.A.

The exam is questioning one's knowledge of the Code and the demand factors that apply in most cases. The math is simple. The calculation question must be read carefully, reading each word as words direct you to where to turn in the Code book.

As you work in this workbook to solve the calculations please show your work. This is very important as it will help you to retain what Code section you applied the Code rules from.

Don't memorize answers, as the numbers will change in the questions. The calculation questions come from categories such as motors, transformers, service sizing, fault-currents, impedance, voltage drop, etc. as shown in this book.

You can't work too many calculation questions prior to an examination. The key to passing is repetition by working and solving questions from all the categories of questioning.

I will personally solve the 101 calculation questions asked in this book and show you the key to follow.

CONTENTS

	PAGE
Alternator, Poles Required	19
Ampacity	33, 48, 63, 71, 78
Box Sizing	13, 30, 41, 69, 81
Branch Circuit, Fluorescent Lights	15
Branch Circuit, Office	8
Branch Circuit, Sizing	28
Cable Tray Size	47
Commercial, Cooking Equipment	36, 90
Commercial, Service	49, 95, 97, 99, 101
Conductors, THHN Size	3
Conduit Size	18
Dryer, Clothes	53
Dwelling, Service Size	42, 91
Fault-Current	23
Freight Elevator	2
Grounding Electrode Size	22, 40
Gutter, Minimum Size	10
Kvars	80
KWH Meter Reading-Cost	65, 67
Load Balancing	56, 79
Marina Receptacle Demand	84
Mobile Home Park Demand	9
Motor Control	52, 75, 82
Motor, Hermetic Fuse Size	17
Motor, Sizing	29, 35, 39, 45, 54, 66, 100
Neutral Load	64, 87, 92, 98
Office, Receptacle Demand	43
Optional Demand, Apartments	61
Power Factor	20, 44
Ranges, 1ø on 3ø System	7
Ranges, Household	14, 58, 73, 93
Resistance-Theory Questions	12, 21, 26, 34, 38, 50, 59, 60, 68, 76, 83, 88
School, Optional Demand	11, 51
Transformer, 3ø, Current	4, 6, 27, 37, 70, 72, 89, 96
Transformer, 3ø, Impedance	1, 55, 57
Transformer, Auto	31, 62
Transformer, Connections	24, 25
Transformer, Ratio	74, 86
Vars	5
Voltage Drop	32, 94
Welder, Sizing	46, 77, 85
Wireway Size	16

CALCULATIONS II

Question: #1

A 167 kva, three-phase, 208/120 volt transformer with 2% impedance.

The maximum fault current available at the secondary terminals of the transformer is ____ amperes.

(a) 25,000   (b) 23,200   (c) 22,300   (d) 464

Show your work with answer choice:

CALCULATIONS II

Question: #2

A 25 hp, 208 volt, three-phase, 15 minute rated motor is used on a freight elevator.

What size branch circuit conductors are required? Use TW copper conductors.

(a) #2   (b) #4   (c) #1   (d) #3

Show your work with answer choice:

CALCULATIONS II

Question: #3

Determine the size of conductors required to supply a 75 amp non-motor load. The device terminations are rated at 75°C. Use copper conductors, conductor insulation type is THHN.

(a) #6   (b) #3   (c) #4   (d) none of these

Show your work with answer choice:

CALCULATIONS II

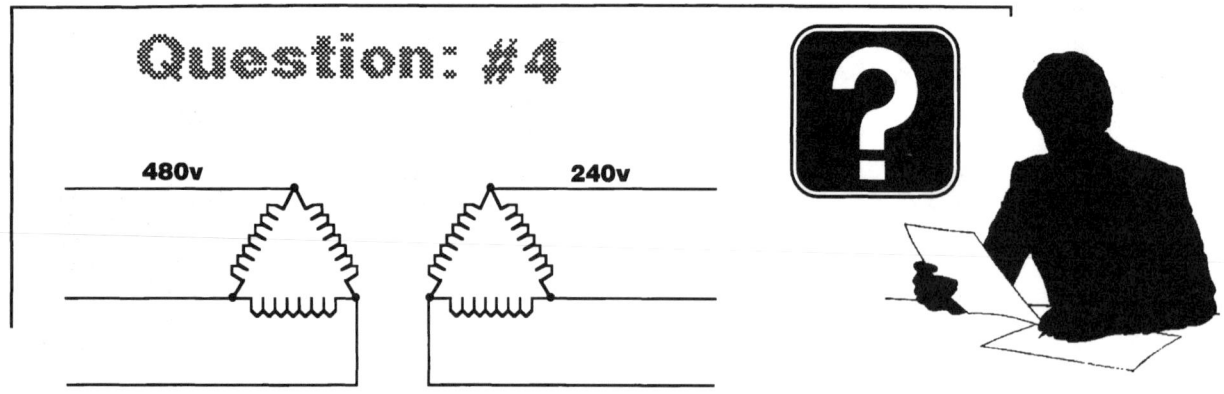

Question: #4

Calculate the primary current of a transformer that has a secondary current of 75 amperes, secondary voltage of 240, 3ø and a primary voltage of 480, 3ø.

(a) 37.5 amps   (b) 150 amps   (c) 130 amps   (d) 40 amps

Show your work with answer choice:

Question: #5

A 240 volt single-phase motor has a current flow of 10 amps. A wattmeter reads 1630 watts. To raise the power factor to 95%, how many vars are required?

Vars _____.

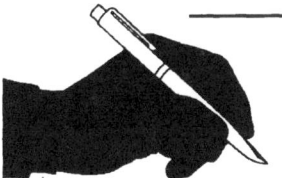

Show your work

CALCULATIONS II

Question: #6

What is the current flow in amps for the neutral of a wye three-phase system given the three individual line currents flowing are: L1 = 60 amps, L2 = 70 amps, and L3 = 80 amps.

(a) 17.32 amps (b) 25.46 amps (c) 60 amps (d) 30 amps

Show your work with answer choice:

Question: #7

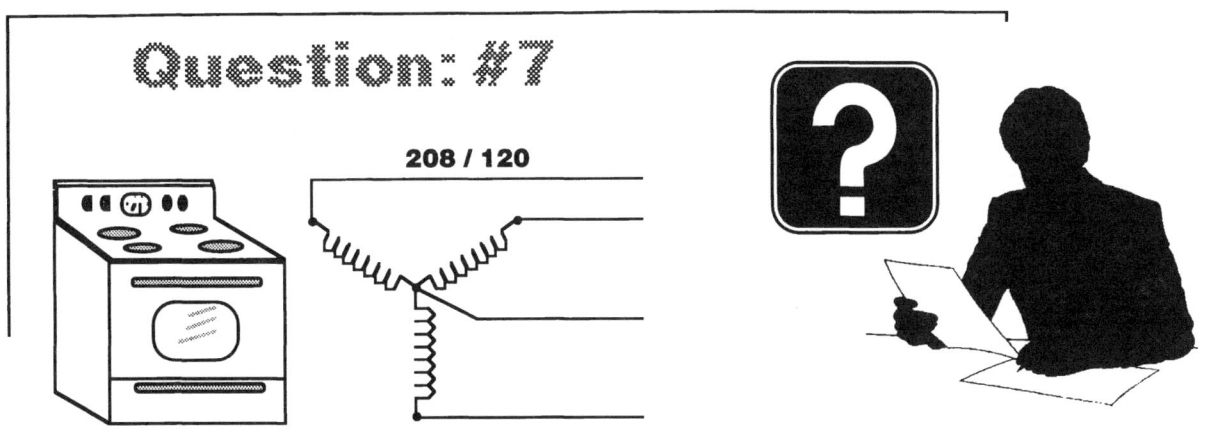

Thirty single-phase household ranges rated at 12kw each are supplied by a 3ø, 4-wire, 120/208v feeder. What is the demand on the feeder for these ranges?

(a) 360 kw (b) 35 kw (c) 45 kw (d) 86.4 kw

Show your work with answer choice:

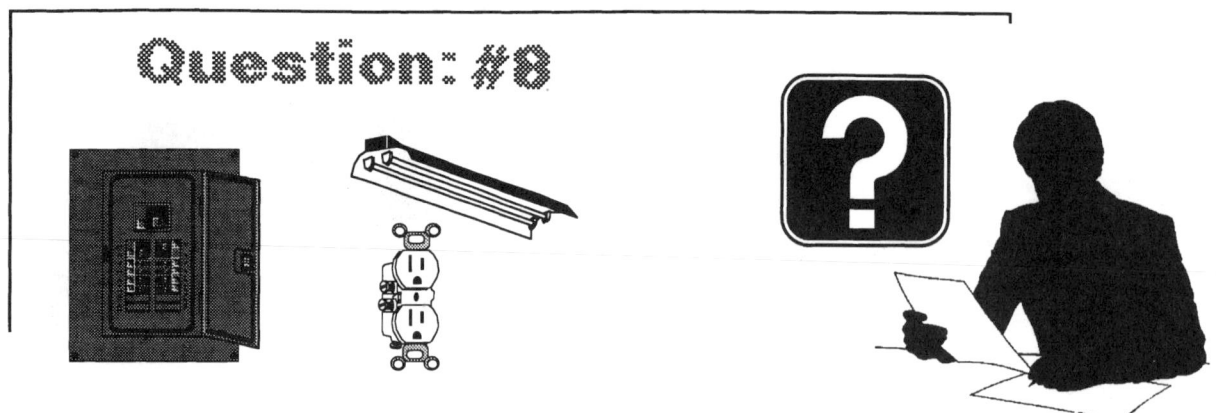

Question: #8

How many 20 amp, 120 volt, 2-wire circuits are needed for a 40,000 sq.ft. office building with an unknown receptacle load?

(a) 40   (b) 60   (c) 90   (d) 120

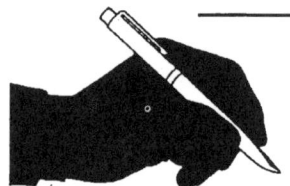

Show your work with answer choice:

CALCULATIONS II

Question: #9

A mobile home park has 25 mobile home sites which would require a minimum demand of ____ kva.

(a) 50   (b) 65   (c) 96   (d) 100

Show your work with answer choice:

Question: #10

Gutters are manufactured in sizes 4" x 4", 6" x 6", 8" x 8", 10" x 10", and 12" x 12" in lengths of one foot up to 10 feet.

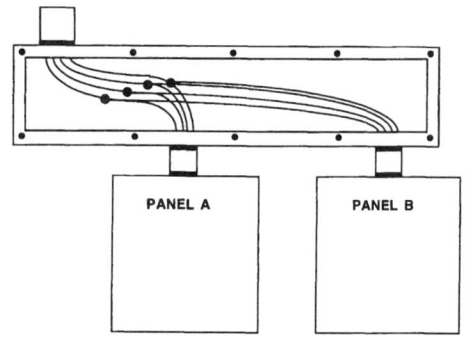

The service has 4 - #300 THHN conductors which are tapped to 8 - #3/0 THHN conductors.

What is the minimum size gutter allowed by the Code?

(a) 4" x 4"   (b) 6" x 6"   (c) 8" x 8"   (d) 10" x 10"

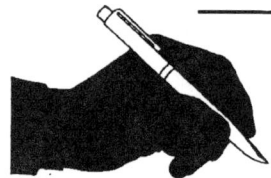

Show your work with answer choice:

CALCULATIONS II

Question: #11

What is the demand for a 25,000 sq.ft. school building with a total connected load of 750 kva? Use the optional method of calculation.

(a) 30 kva   (b) 456,250va   (c) 1500 amps   (d) none of these

Show your work with answer choice:

Question: #12

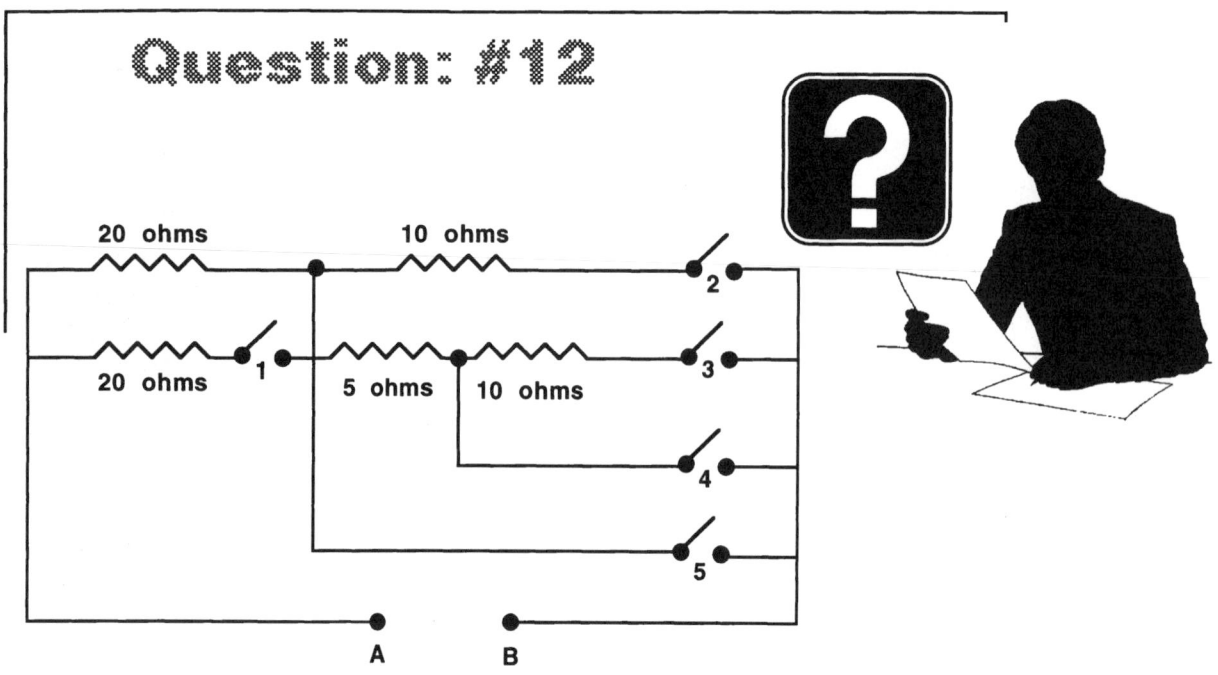

The resistance is ____ ohms when switches 1 and 3 are closed.

(a) 25   (b) 35   (c) 55   (d) 65

Show your work with answer choice:

Question: #13

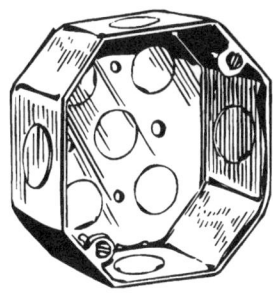

What size octagon box is required for 4 - #12 and 3 - #14 conductors?

(a) 1 1/4"   (b) 1 1/2"   (c) 2"   (d) 2 1/8"

Show your work with answer choice:

CALCULATIONS II

Question: #14

In a custom house, the demand load (load applied for service calculation) for a 12 kw range and a 4 kw oven is _____ kw.

(a) 11   (b) 11.2   (c) 12   (d) 16

Show your work with answer choice:

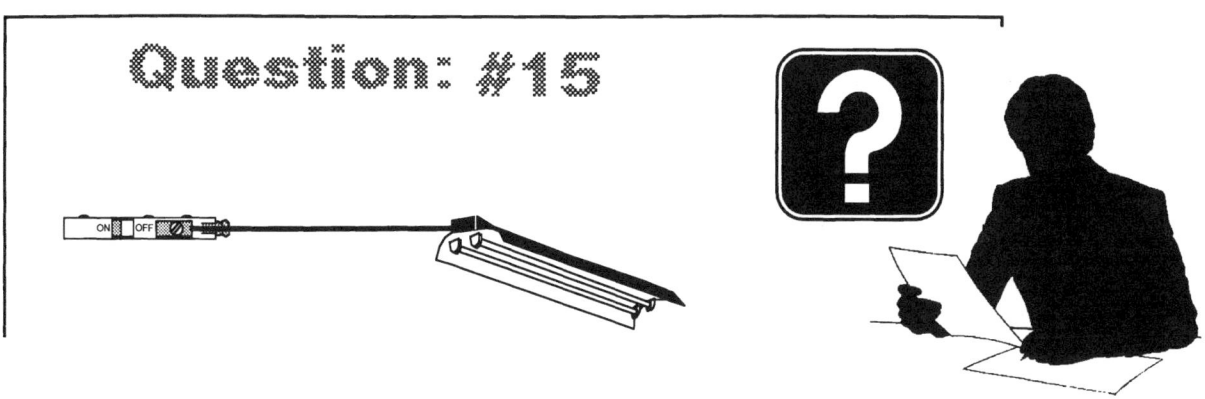

Question: #15

Forty-six 250 watt, 120 volt, flourescent lights are connected to separate 20-amp circuits, balanced on a 3ø, 208/120v system. How many circuits are required, if these lights are on for more than three hours continuously?

(a) 8   (b) 7   (c) 5   (d) 4

Show your work with answer choice:

Question #16

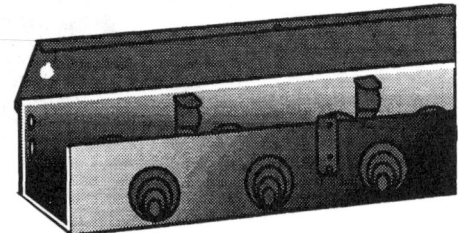

What is the area of square inch of the following THW conductors to be installed in a wireway? 4 - #2/0, 6 - #6, and 20 - #12.

(a) 1.1742   (b) 2.0052   (c) 3.4132   (d) 1.1990

Show your work with answer choice:

Question #17

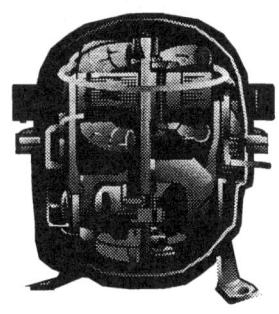 

Determine the size of the time-delay fuses required for a 42 amp hermetic compressor, 240 volts, single-phase.

(a) 75 amps   (b) 80 amps   (c) 85 amps   (d) 90 amps

Show your work with answer choice:

Question #18

What is the minimum size rigid metal conduit required for the following conductors?

3 - #500 kcmil XHHW
3 - #8 THW
6 - #14 THHN

(a) 2 1/2" (b) 3" (c) 3 1/2" (d) 4"

Show your work with answer choice:

Question #19

How many poles must a large alternator turning at 1800 rpm have to produce a 60 Hz frequency?

(a) 2   (b) 4   (c) 6   (d) 8

Show your work with answer choice:

Question #20

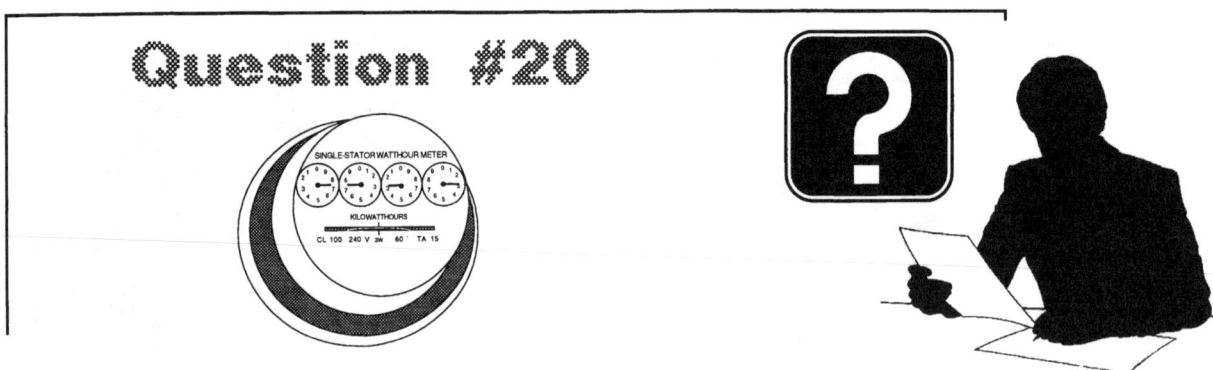

You have a load of 7500 kw, @ 75% PF, @ 12,470 volts, 3 ø. The current drawn would be closest to _____ amperes.

(a) 5   (b) 50   (c) 500   (d) 5000

Show your work with answer choice:

Question #21

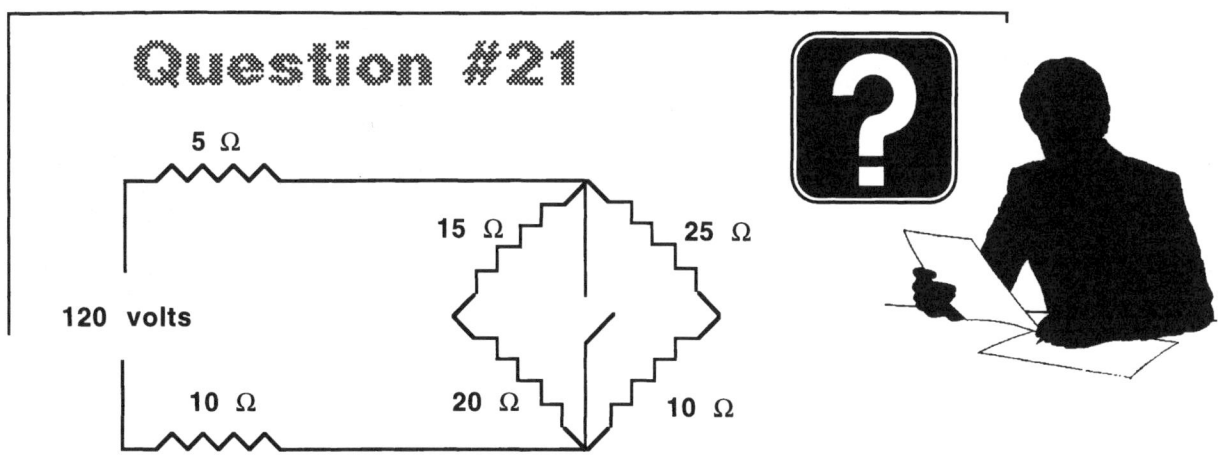

What is the total resistance of this 120 volt circuit with the switch open as shown?

(a) 85 ohms   (b) 17.5 ohms   (c) 32.5 ohms   (d) 95 ohms

Show your work with answer choice:

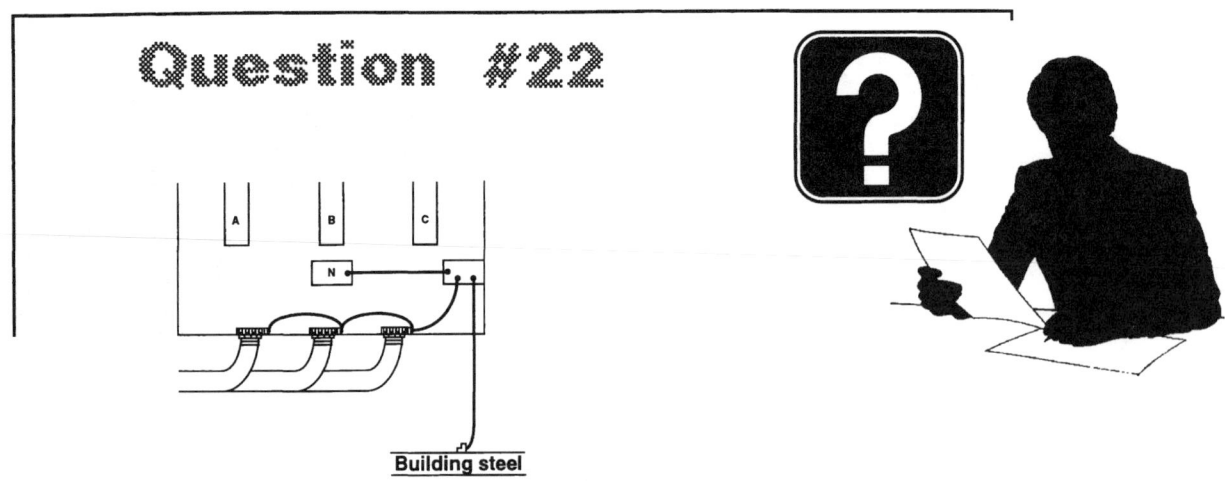

The service entrance contains 3 - #500 kcmil copper conductors paralleled per phase. What size copper grounding electrode conductor is required?

(a) #1   (b) #1/0   (c) #3/0   (d) #4/0

Show your work with answer choice:

Question #23

Calculate the maximum fault current available at the secondary terminals of a 750 kva transformer, 3 ø, 240v, with 5 1/2% impedance.

(a) 24,600 amps   (b) 32,800 amps   (c) 180,400 amps   (d) 1804 amps

Show your work with answer choice:

Question #24

Complete the 3ø transformer diagram below to provide 480 volt delta primary and 208/120 volt wye connected secondary. Label voltages, transformer terminals (H1,H2,X1,X2) and neutral.

Show your work:

Question #25

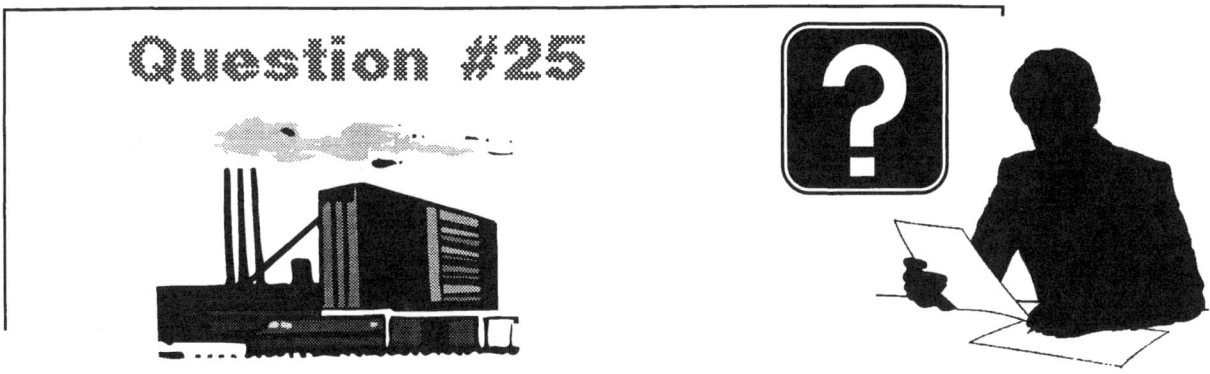

An industrial plant is served at 12,470 volts, 3ø, 4-wire. A new 900 kva load is to be served at 4160 volts, 3ø, 4-wire. Three 300 kva, 1ø transformers rated @ 14,400/7200 - 4800/2400 volts are to be used. Draw the connections on the diagram below. Label all voltages and terminals.

Show your work:

Question #26

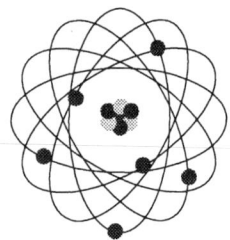

In the circuit shown below, the ammeter would read ____ amps.

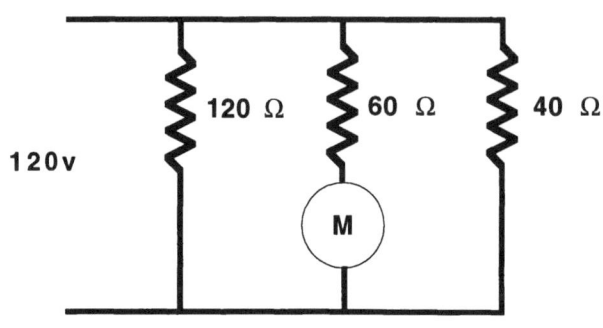

(a) 2 (b) 6 (c) 60 (b) .545

Show your work with answer choice:

Question #27

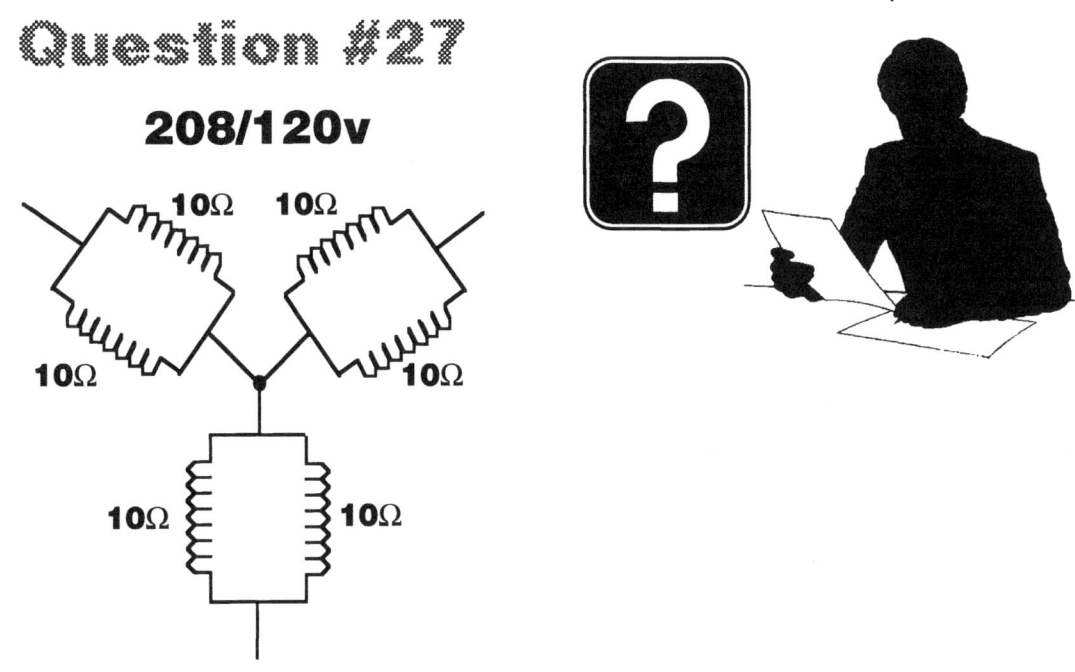

What is the phase amps in the three-phase wye 208/120v transformer?

(a) 20.8 amps   (b) 12 amps   (c) 24 amps   (d) 5 amps

Show your work with answer choice:

Question #28

A single 1500 watt cord and plug connected load on a 120 volt branch circuit would draw ____ amps, this requires a number ____ wire and ____ circuit breaker for the branch circuit.

(a) 8 - #14 - 15 amp

(b) 10.5 - #14 - 15 amp

(c) 12.5 - #14 - 15 amp

(d) 12.5 - #12 - 20 amp

Show your work with answer choice:

Question #29

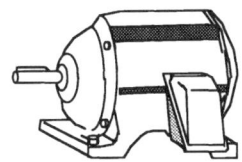

What size inverse time circuit breaker should be selected for a branch circuit to a 3ø, 3hp, 208v motor?

(a) 15 amps (b) 20 amps (c) 25 amps (d) 35 amps

Show your work with answer choice:

Question #30

What is the cubic inch capacity required for a device box containing one duplex receptacle, cable clamps, and two #12-2 with ground nonmetallic sheathed cables (romex)?

(a) 13.5 cu.in.   (b) 15.75 cu.in.   (c) 16 cu.in.   (d) 18 cu.in.

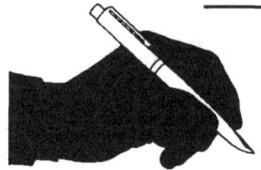

Show your work with answer choice:

CALCULATIONS II

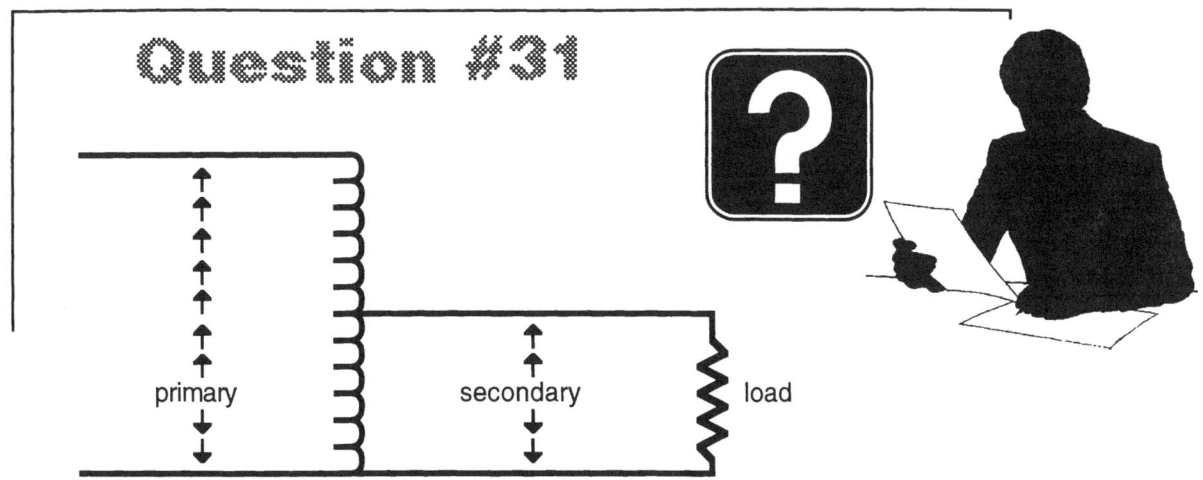

Question #31

An autotransformer with a 2 to 1 ratio has a primary voltage of 100 volts. The secondary load is 5 ohms. The current in the load is approximately ____ amps.

(a) 5 (b) 10 (c) 15 (d) 20

Show your work with answer choice:

Question #32

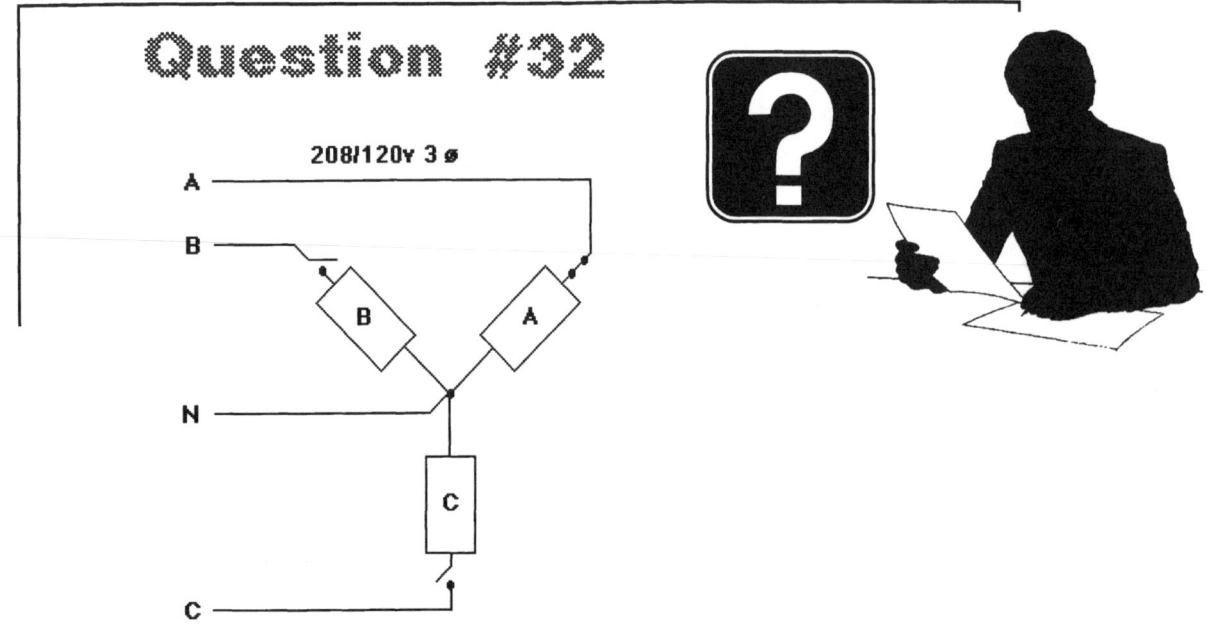

If load "A" is 9.6 kva and the voltage drop is not to exceed 3%, what size copper THW conductor is necessary if the load is located 90 feet from the source?

(a) #2   (b) #3   (c) #4   (d) none of these

Show your work with answer choice:

Question #33

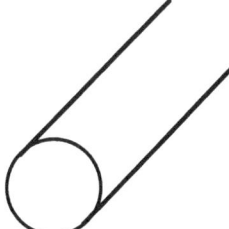

The ampacity of a 2" round bus bar is approximately ____ amperes.

(a) 3572   (b) 3142   (c) 2836   (d) 1741

Show your work with answer choice:

Question #34

In the circuit shown below all three resistors have the same ohmic value, the resistance of R2 would be ____ ohms.

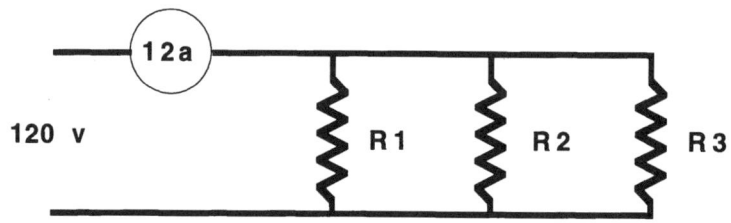

(a) 10   (b) 20   (c) 30   (d) 40

Show your work with answer choice:

Question #35

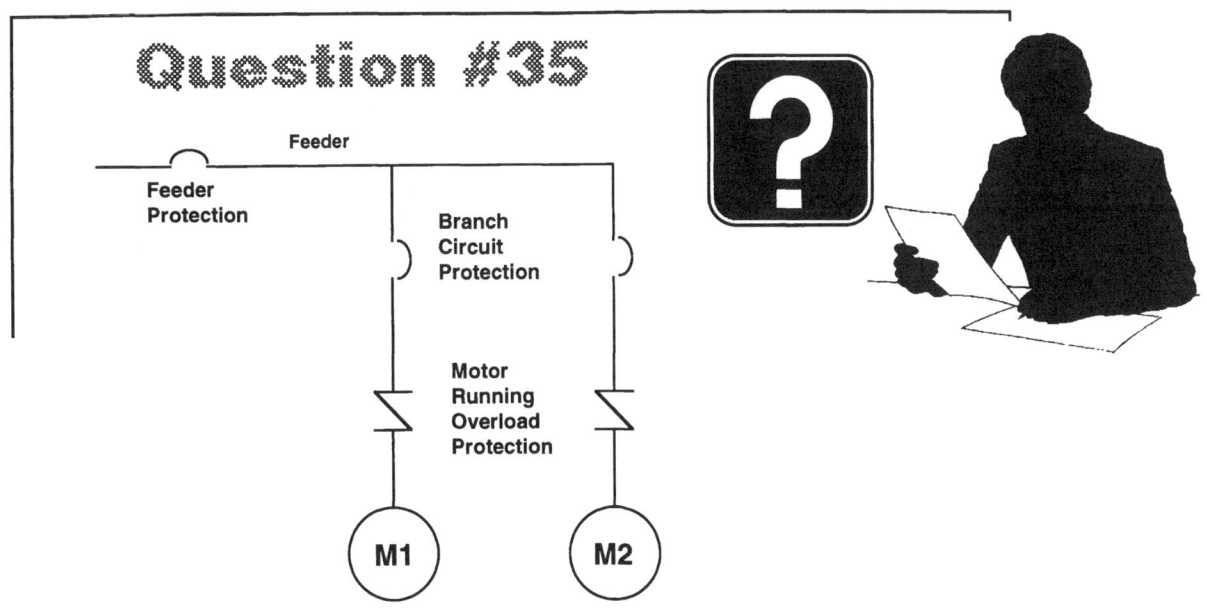

If M1 is a 1 ø 3 hp motor on 208v and M2 is a 3ø 5 hp motor on 208v, what is the largest inverse time breaker allowed for the feeder protection?

(a) 40 amp   (b) 50 amp   (c) 60 amp   (d) none of these

Show your work with answer choice:

Question #36

Given: The following 480 volt, 3-phase, 3-wire, intermittent use equipment is in a commercial kitchen:

2 - 5,000 watt water heaters
4 - 3,000 watt fryers
2 - 6,000 watt ovens

Each ungrounded conductor in the feeder circuit for this kitchen equipment must be sized to carry a minimum computed load of ____ .

(a) 13 amps   (b) 18 amps   (c) 21 amps   (d) 27 amps

Show your work with answer choice:

Question #37

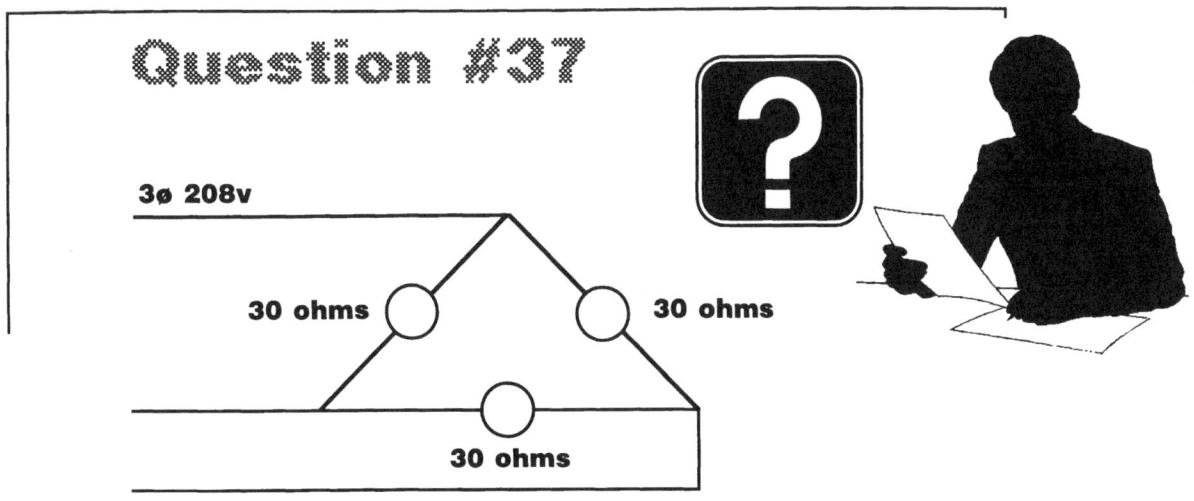

IL = ____.

(a) 6.9 (b) 13.9 (c) 12 (d) 20.7

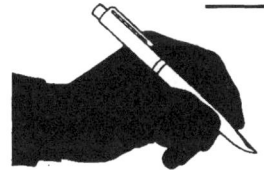

Show your work with answer choice:

CALCULATIONS II

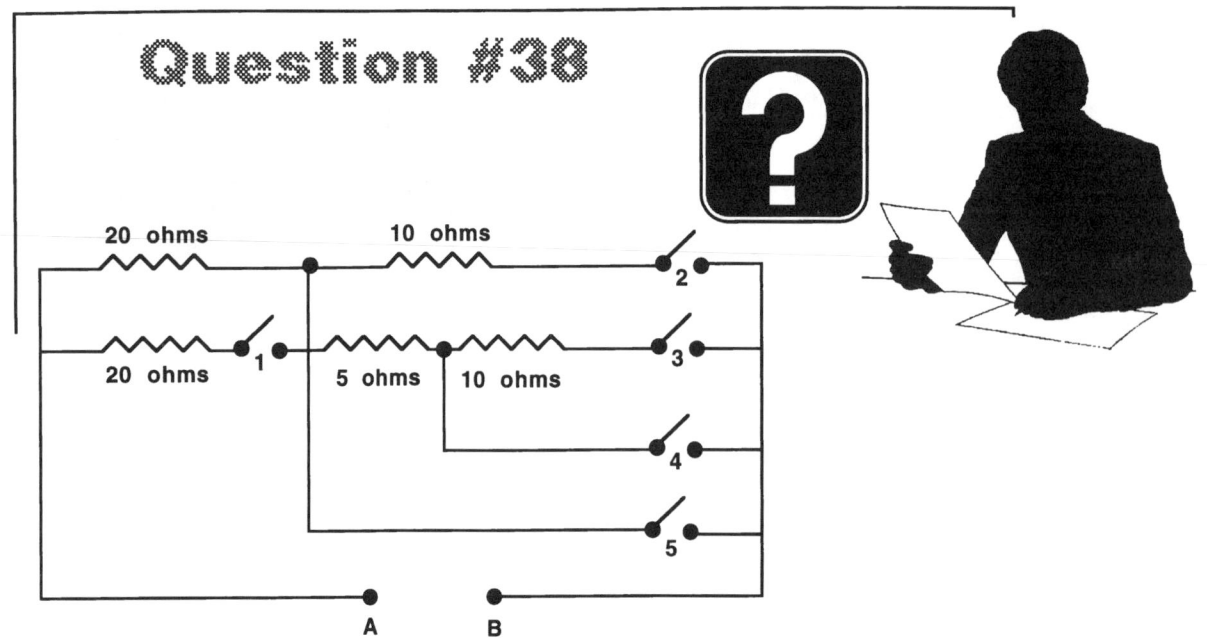

Resistance A-B is lowest when switches ____ are open.

(a) 1 through 5   (b) 2 through 5   (c) 2, 3, and 4   (d) 3, 4, and 5

Show your work with answer choice:

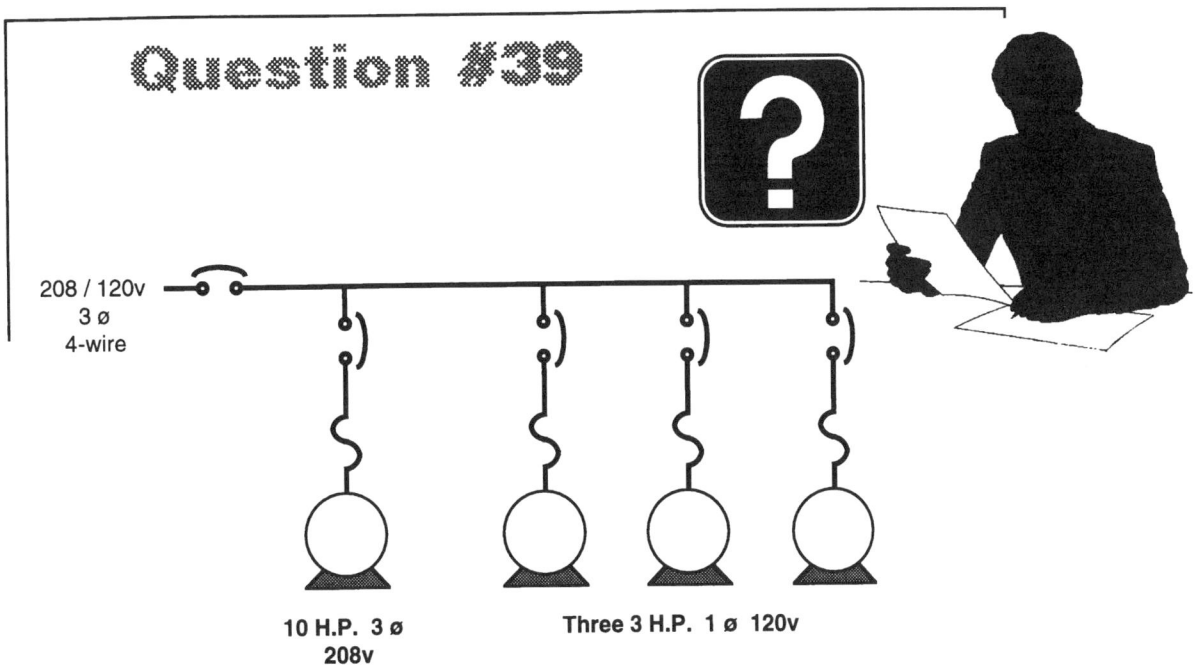

What size feeder conductor is required using TW insulation?

(a) #2/0   (b) #1/0   (c) #3   (d) #4

Show your work with answer choice:

CALCULATIONS II

Question #40

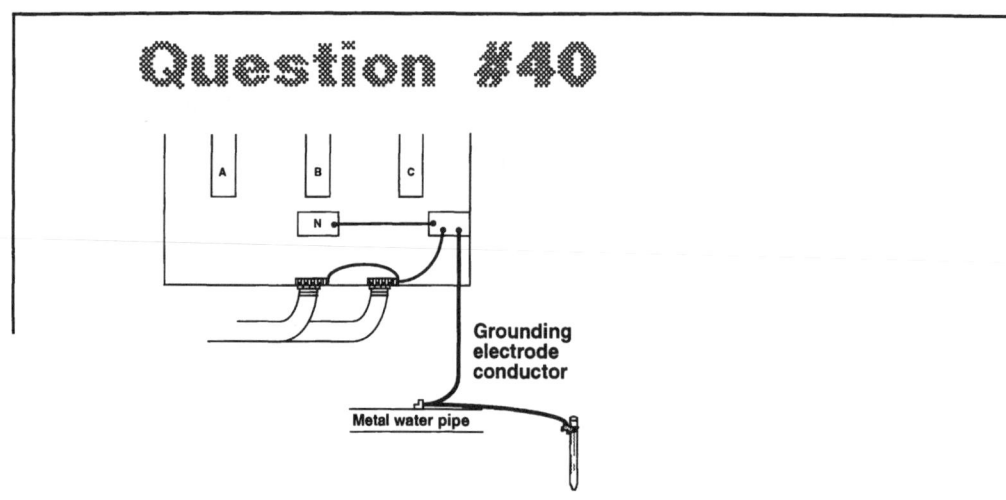

A three-phase service has 2 - #600 THW copper conductors paralleled per phase. What size copper grounding electrode conductor is required from the service equipment to the water pipe?
#_____.

What size copper conductor is required from the water pipe to the ground rod?
#_____.

Show your work with answer choice:

CALCULATIONS II

Question #41

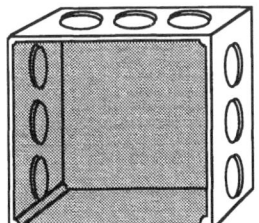

What size square box is required for the following conductors?

4 - #14 THW, 4 - #12 THW, 5 - #10 THW, and 3 - #8 THW

(a) 4" x 2 1/8"   (b) 4 11/16" x 1 1/4"

(c) 4 11/16" x 1 1/2"   (d) 4 11/16" x 2 1/8"

Show your work with answer choice:

What is the demand load on the service for a 1500 sq.ft. dwelling unit with the following: (use general method) 6 kw water heater, 14 kw range, 5.5 kw dryer, and a 10 amp washing machine 120v.

(a) 15,000 - 20,000 va
(b) 20,000 - 26,000 va
(c) 26,000 - 28,000 va
(d) 28,000 - 35,000 va

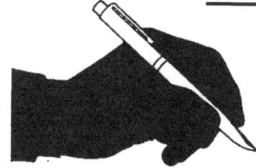

Show your work with answer choice:

Question #43

An office building contains 250 receptacles rated @ 20a, 120v. What is the demand load in kva for these receptacles?

(a) 25-100   (b) 100-200   (c) 200-300   (d) 300-500

Show your work with answer choice:

CALCULATIONS II

Question #44

You have a load of 3320 kva @ 80% power factor @ 480 volts, 3ø. The current would be closest to ____ amps.

(a) 5   (b) 50   (c) 500   (d) 5000

Show your work with answer choice:

CALCULATIONS II

Question #45

What is the LRC (locked rotor current) for a motor with a F.L.C. of 30 amps?

(a) 150 amps   (b) 60 amps   (c) 180 amps   (d) 37.5 amps

Show your work with answer choice:

CALCULATIONS II

Question #46

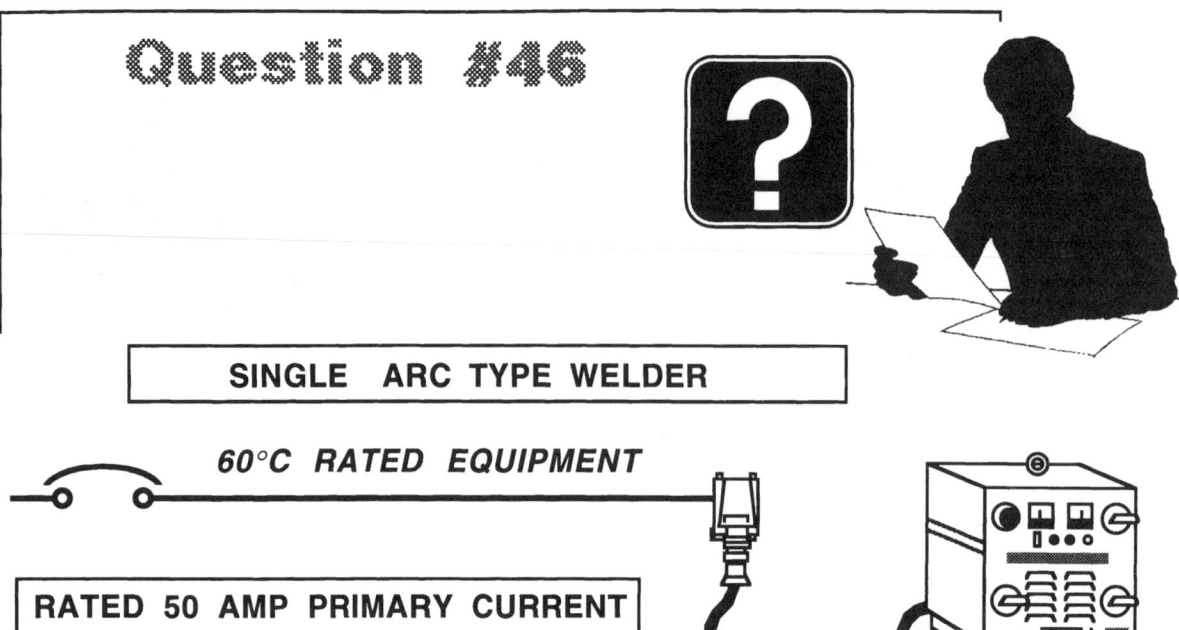

SINGLE ARC TYPE WELDER

60°C RATED EQUIPMENT

RATED 50 AMP PRIMARY CURRENT
240 VOLT 1ø   DUTY CYCLE 80%

The welder would require a ____ circuit breaker.

(a) 50 amp   (b) 60 amp   (c) 70 amp   (d) 100 amp

Show your work with answer choice:

Question #47

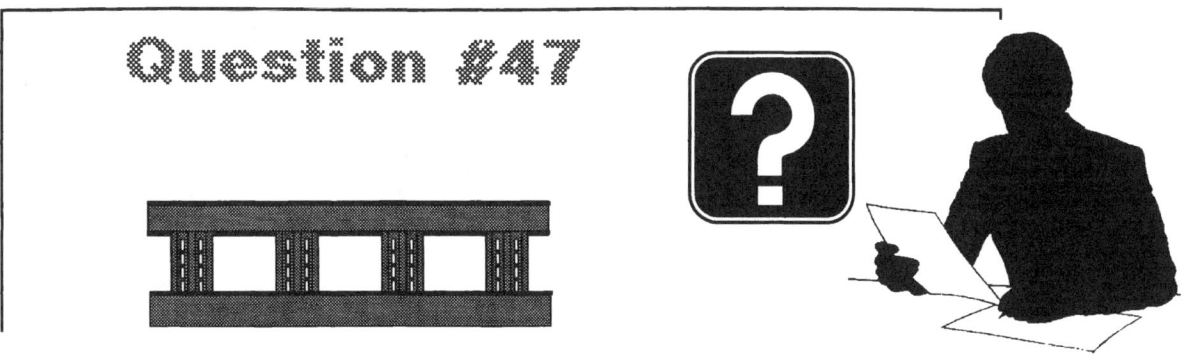

The minimum square inches required for a cable tray containing 8 - #750 kcmil and 6 - #1000 kcmil THW conductors would be ____.

(a) 17.61   (b) 18.192   (c) 19.18   (d) 20.11

Show your work with answer choice:

CALCULATIONS II

The ampacity of a copper #10 THWN-2 is ___ when there are three conductors in the conduit and the ambient temperature is 70°F.

(a) 30   (b) 32.4   (c) 35   (d) 41.6

CALCULATIONS II

If each unit of a 150 unit motel was 15' x 30' and each unit wing had 4,000 sq.ft. of halls and corridors, the feeder demand lighting load for each of the 3 wings would be _____ watts.

(a) 47,000   (b) 22,500   (c) 22,000   (d) 20,800

Show your work with answer choice:

Question #50

In the circuit shown below, the source voltage would be ____ volts.

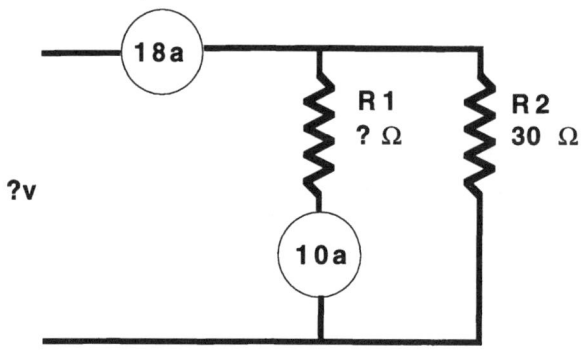

(a) 58 (b) 120 (c) 90 (d) 240

Show your work with answer choice:

CALCULATIONS II

Question #51

What is the demand for a 20,000 sq.ft. school building with a total connected load of 680 kva? Use the optional method of calculation.

(a) 680 kva (b) 385 kva (c) 170 kva (d) none of these

Show your work with answer choice:

Question #52

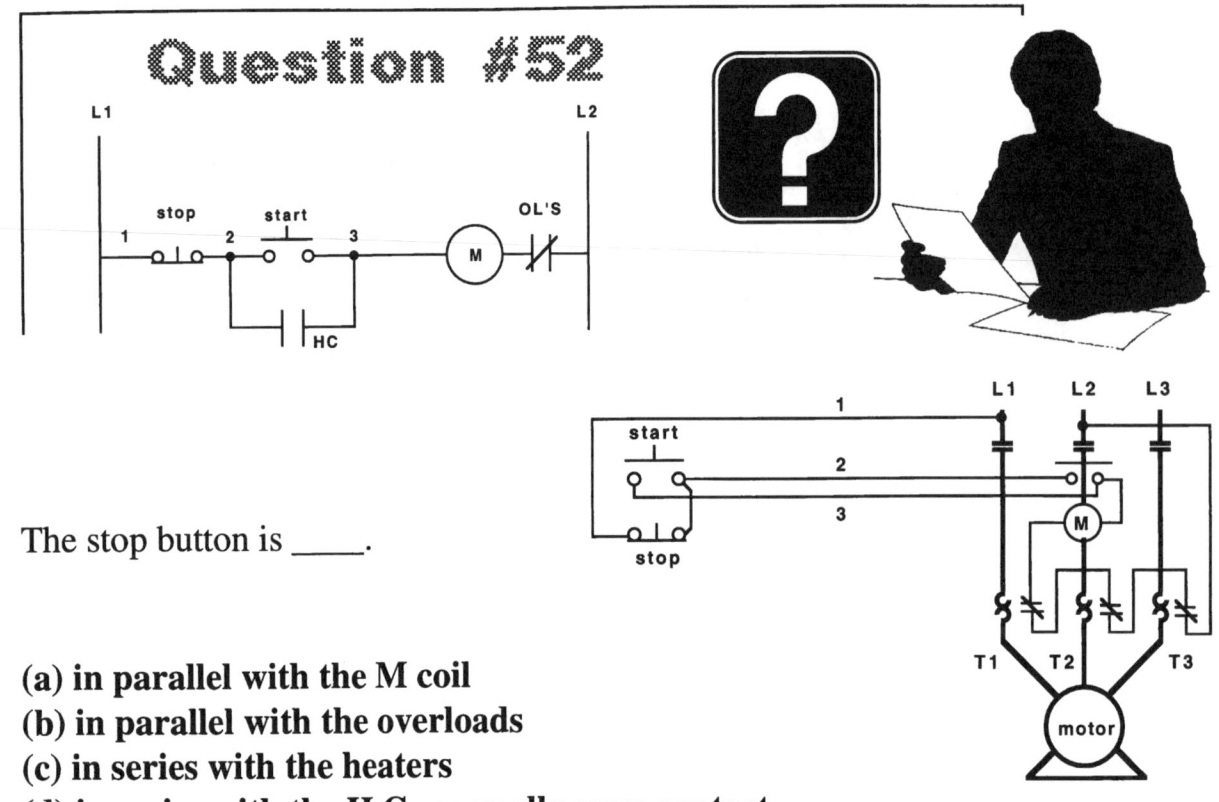

The stop button is ____.

(a) in parallel with the M coil
(b) in parallel with the overloads
(c) in series with the heaters
(d) in series with the H.C. normally open contact

Show your work with answer choice:

An apartment has 200 units. Each unit has its own household electric dryer rated 4 kw. The load to be added to the service calculation after demand factors are included would be ____. (Using the standard method.)

(a) 200 kw   (b) 250 kw   (c) 800 kw   (d) 500 kw

Show your work with answer choice:

CALCULATIONS II

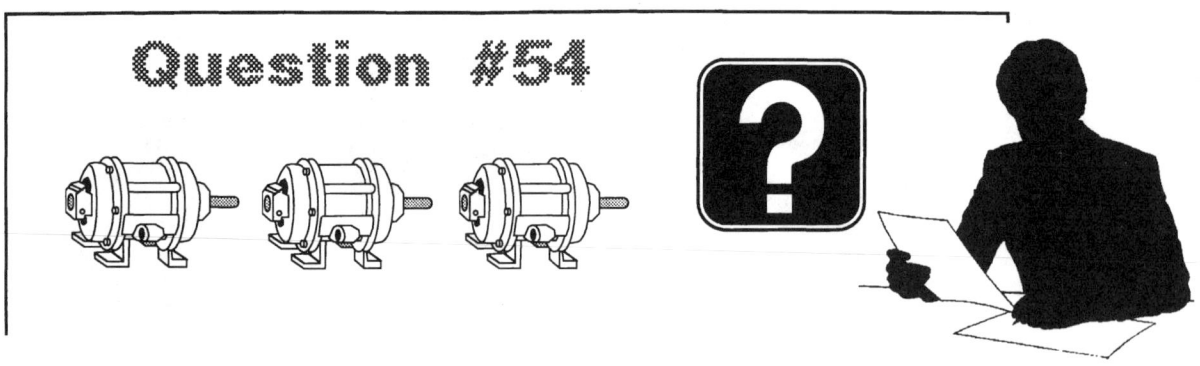

Motors of 1/3, 1/4, and 1/8 hp are connected in parallel. The three motors deliver a total of ____ horsepower.

(a) 1   (b) 7/8   (c) 17/24   (d) .07

Show your work with answer choice:

The available short-circuit current from this transformer would be ____ amperes.

(a) 15,000 - 25,000   (b) 26,000 - 50,000   (c) 51,000 - 75,000   (d) 76,000 - 95,000

Show your work with answer choice:

Balance the following loads on a 208 / 120v 4-wire three-phase secondary.
1 - 6000va 3ø 208v motor
3 - 5000va 1ø 208v heaters
3 -   150va 1ø 120v lights

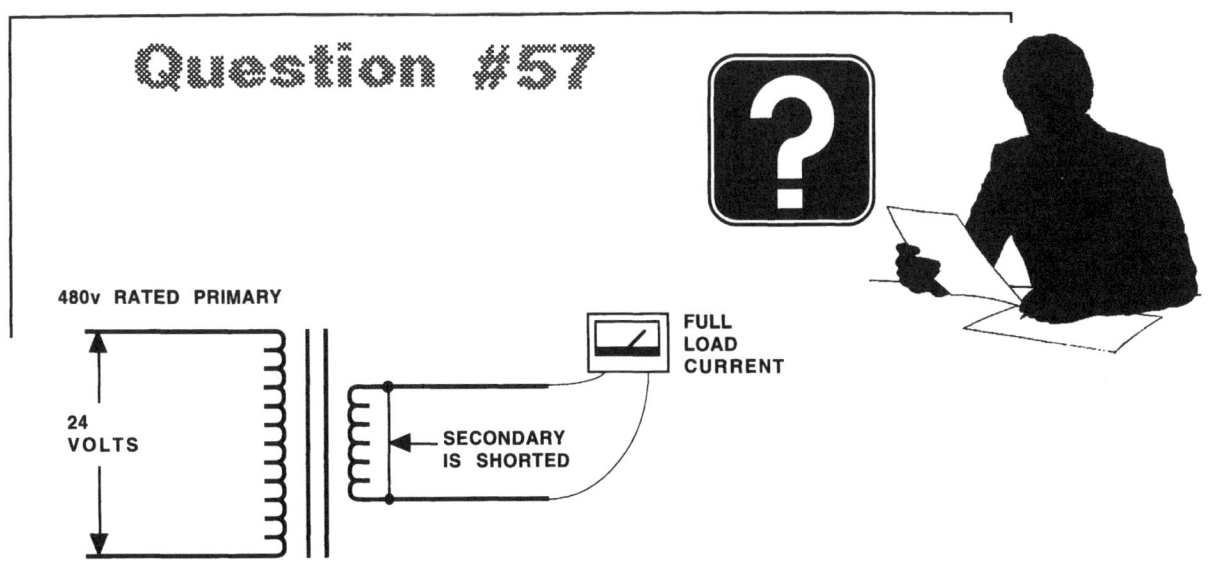

A 480 volt rated primary has 24 volts applied which causes secondary full load current to flow through the shorted secondary. The transformer impedance would be ____.

(a) 20%   (b) 2%   (c) 5%   (d) 95%

Show your work with answer choice:

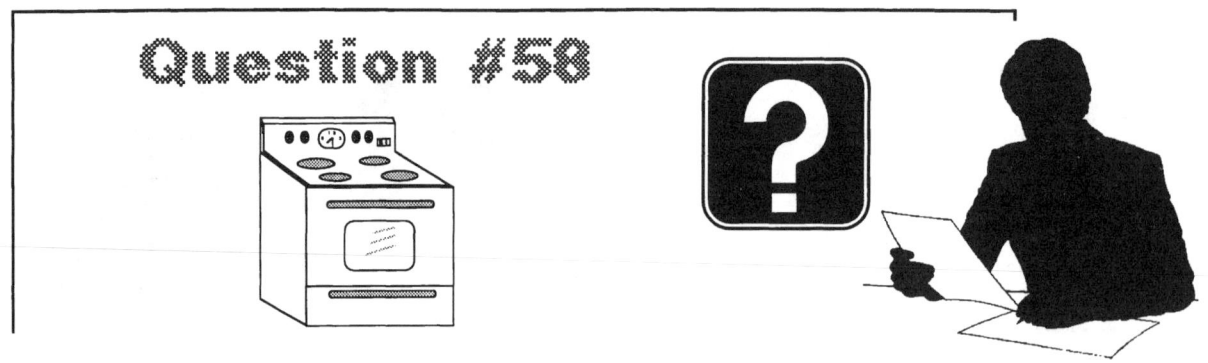

What is the maximum feeder demand for 35 - 8kw ranges in an apartment complex?

(a) 280 kw   (b) 67.2 kw   (c) 61.6 kw   (d) 50 kw

Show your work with answer choice:

Question #59

A nichrome wire having a resistance of 250 ohms per 100 feet is used to wire a toaster. What would be the total length of wire if the toasters total resistance is 10 ohms?

(a) 4 feet   (b) 5 feet   (c) 25 feet   (d) none of these

Show your work with answer choice:

CALCULATIONS II

Question #60

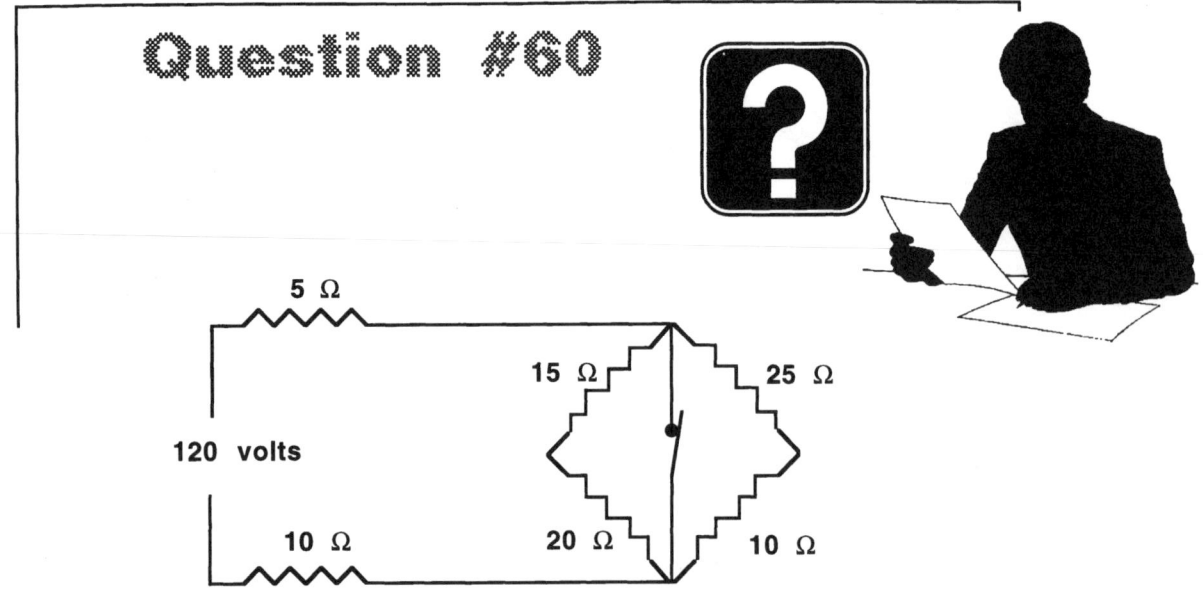

What is the current flowing in this circuit with the switch closed?

(a) 1.41 amps   (b) 1.71 amps   (c) 3.69 amps   (d) 8 amps

Show your work with answer choice:

CALCULATIONS II

Question #61

By the optional method, the service for a 24 unit apartment would be ____ if each apartment was 900 sq.ft. and the service was 230/115v 1ø. Each apartment contains:

6000w of central electrical space heating 230v
3 hp A/C 230v
8 kw range 230/115v
4.5 kw water heater 230v
1.2 kw dishwasher 115v

(a) 100-200 kw   (b) 200-400 kw   (c) 400-600 kw   (d) 600-700 kw

Show your work with answer choice:

Question #62

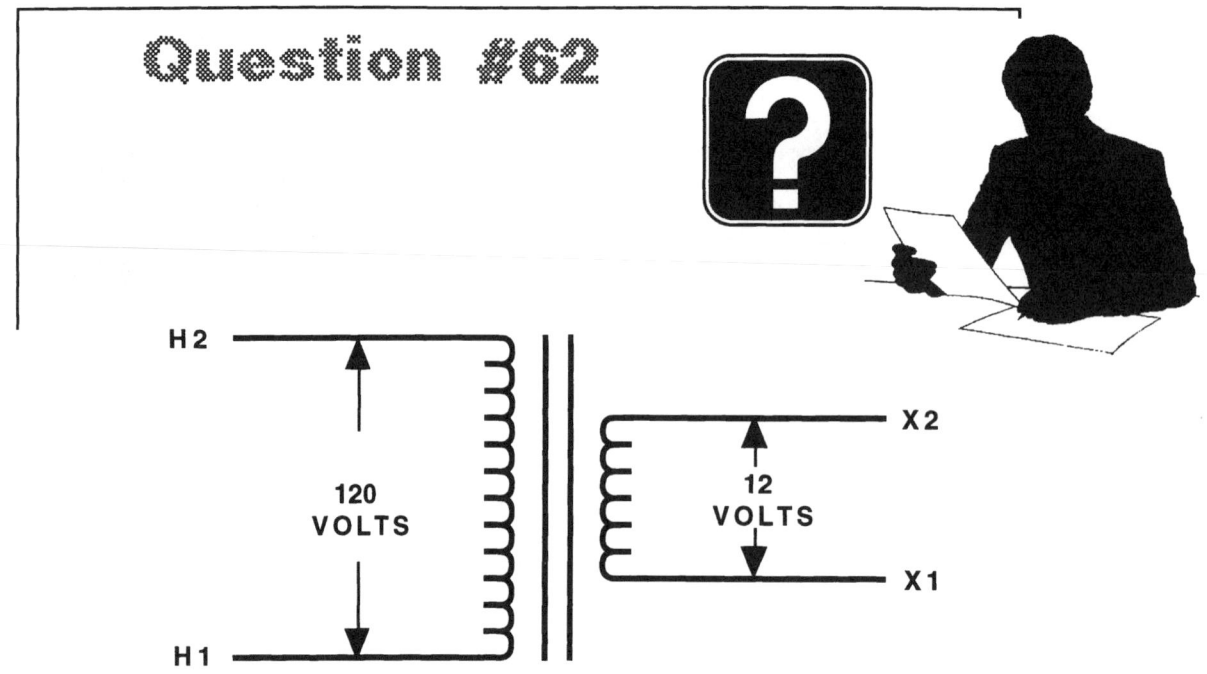

Draw the connections for an autotransformer to deliver 132 volts.

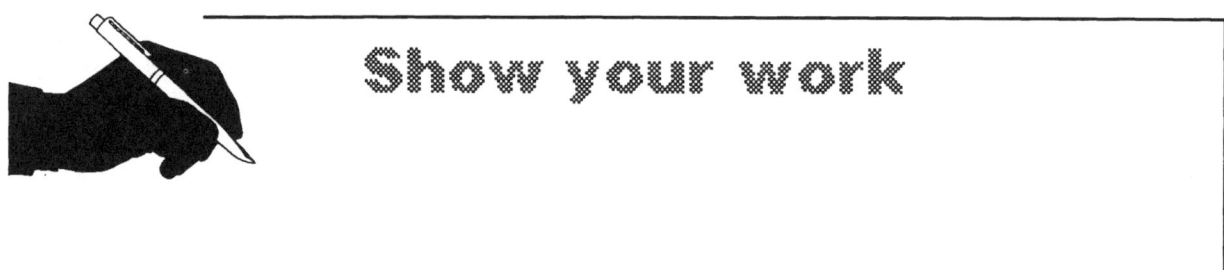

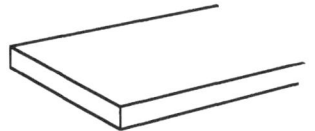

What is the ampacity of a 1/4" x 1 1/2" aluminum bus bar?

(a) 1225 amps   (b) 1750 amps   (c) 461.5 amps   (d) 262.5 amps

Show your work with answer choice:

CALCULATIONS II

Question #64

NEUTRAL

Find the neutral load for the following, the service voltage is 115/230v single-phase.

twelve 800 square foot apartments
twelve 12kw ranges
twenty-four small appliance circuits
twelve laundry circuits

(a) 212 amps (b) 217 amps (c) 252 amps (d) 365 amps

Show your work with answer choice:

Question #65

How many kilowatt hours were used during the month?

_____ hours.

FIRST OF MONTH

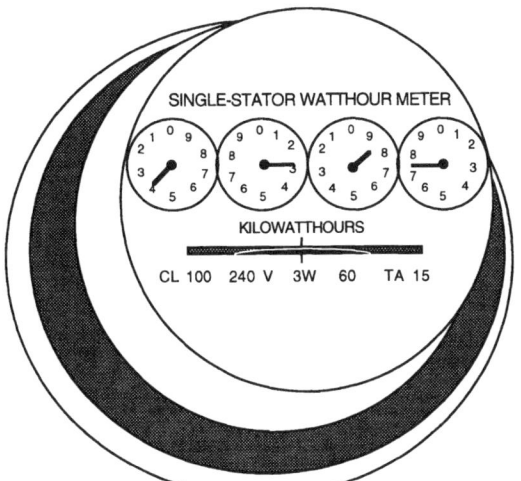

END OF MONTH

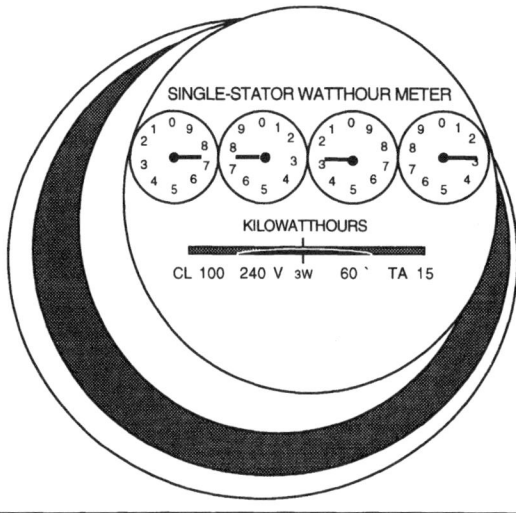

Show your work

Question #66

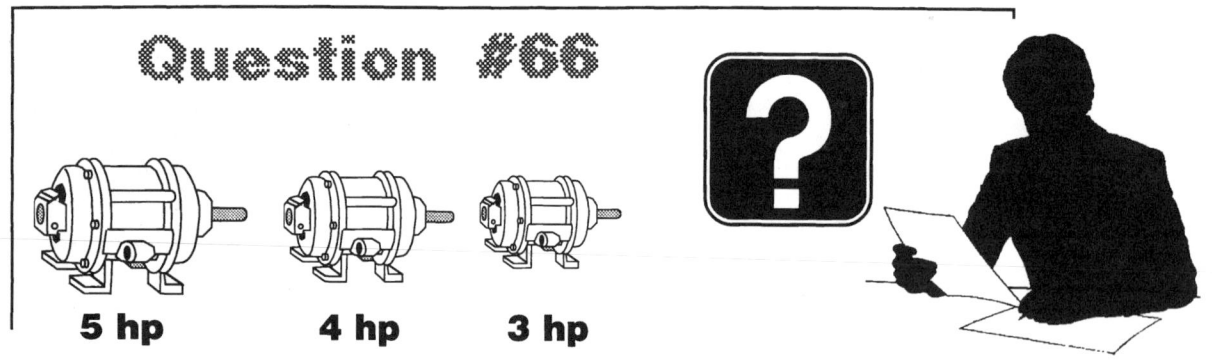

5 hp 4 hp 3 hp

Interpolate the full-load current for a single-phase 230 volt 4 hp motor.

Full-load current _____ **amps.**

Show your work

Question #67

A ceiling fan rated at 120 volts, 2.5 amperes runs continuous for 8 hours each day. What would this fan add to the electric bill over 30 days @ 8 cents a kwh?

Cost $ _____.

Show your work

CALCULATIONS II

Question #68

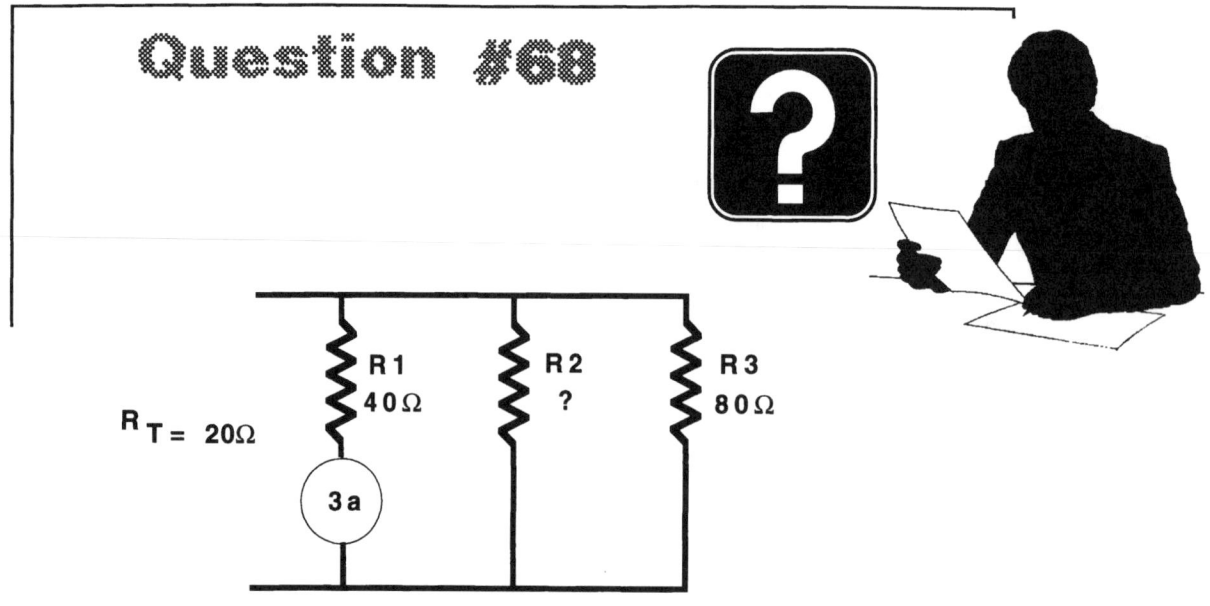

In the circuit shown above, if all three resistors are rated at 250 watts, which resistor or resistors would overheat?

(a) R1   (b) R2   (c) R3   (d) none of them

Show your work with answer choice:

CALCULATIONS II

Question #69

What is the minimum cubic inch allowed for the box shown below? The box contains three cable clamps, two - #12-2 w/grd romex cables to the duplex receptacle and one- #12-2 w/grd romex cable to the single-pole switch.

(a) 34 cu.in.   (b) 31.5 cu.in.   (c) 27 cu.in.   (d) none of these

Show your work with answer choice:

CALCULATIONS II

Question #70

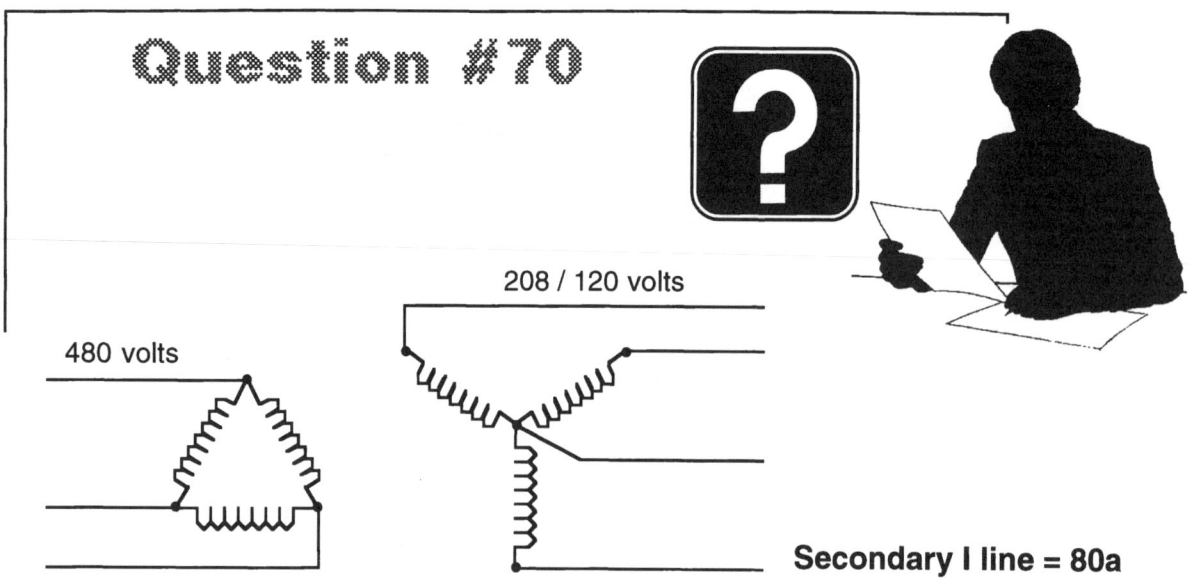

Secondary I line = 80a

What is the primary line current in the transformer shown below?

(a) 20   (b) 34.64   (c) 40   (d) none of these

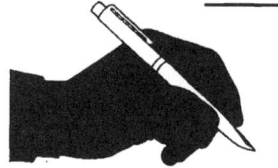

Show your work with answer choice:

CALCULATIONS II

Question #71

What is the ampacity of a #12 THHN copper conductor in an ambient temperature of 122°F with a total of six current-carrying conductors in the conduit?

(a) 19.68 amps (b) 24.6 amps (c) 32.8 amps (d) 20 amps

Show your work with answer choice:

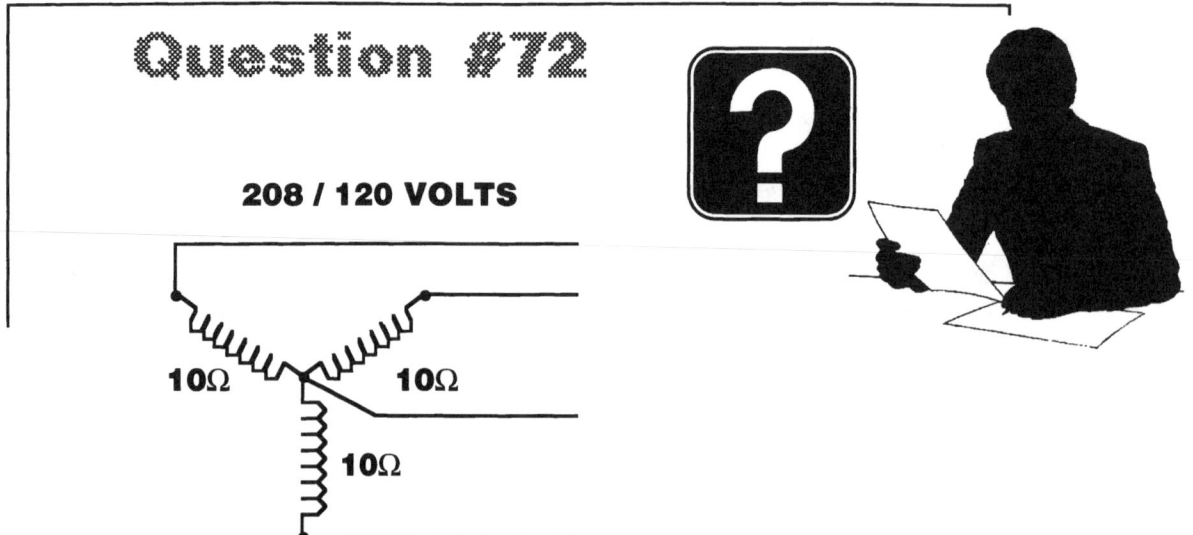

Question #72

208 / 120 VOLTS

What is power dissipated in the resistance of this load?

(a) 720 (b) 1440 (c) 2160 (d) 4320

Show your work with answer choice:

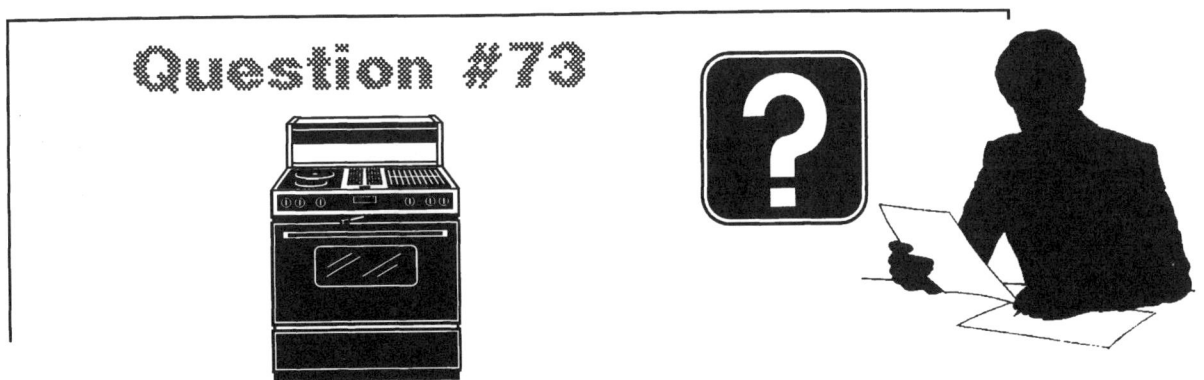

Question #73

What is the feeder demand for ten - 12 kw household ranges 240/120v single-phase connected to a three-phase system?

(a) 120 kw   (b) 40 kw   (c) 25 kw   (d) 23 kw

Show your work with answer choice:

Question #74

A transformer has a 3 to 2 ratio. Its primary winding is designed for 240 volts. The secondary voltage of this transformer is ____ volts.

(a) 115   (b) 120   (c) 160   (d) 215

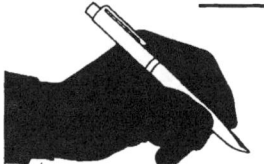

Show your work with answer choice:

Question #75

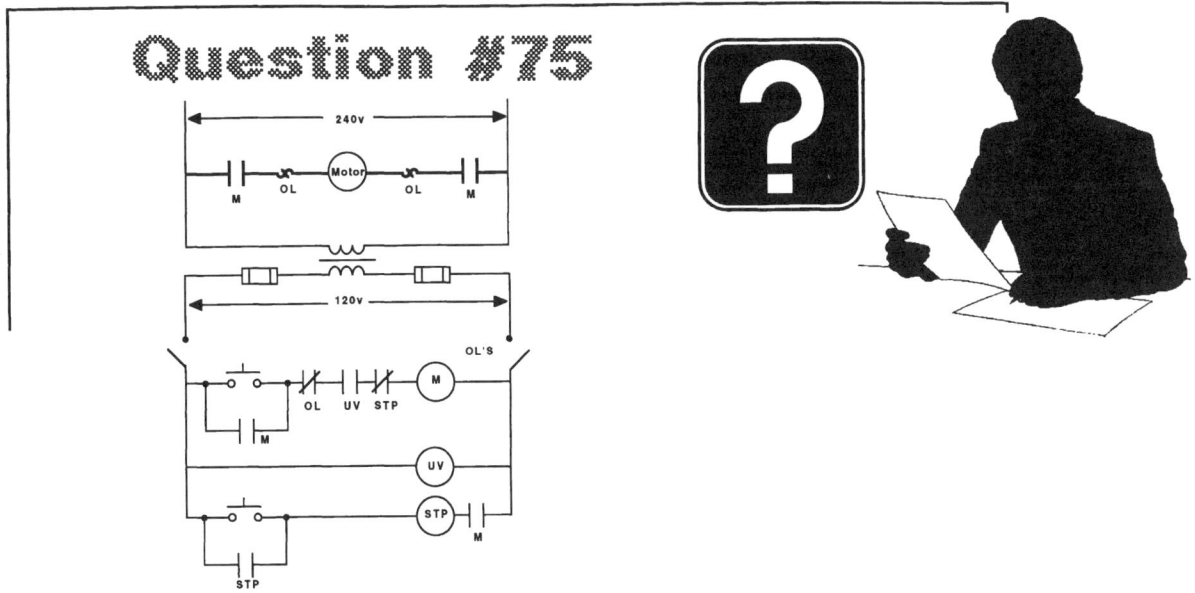

If the fused disconnect were opened, the motor would stop because of all of the following except ____.

(a) under voltage
(b) the "M" coil de-energizing
(c) the "M" main contactors dropping open
(d) the "STP" coil de-energizing

Show your work with answer choice:

If the RMS (effective value) is 480 volts, the maximum (peak) voltage of an AC system would be approximately ____ volts.

(a) 480   (b) 577   (c) 679   (d) 707

Show your work with answer choice:

CALCULATIONS II

Question #77

A single non motor generator welder rated at 50 amps primary current @ 240 volts with a 50% duty cycle would require # ____ THHN conductors. The equipment is rated 60°C.

(a) 10   (b) 8   (c) 6   (d) 4

Show your work with answer choice:

CALCULATIONS II

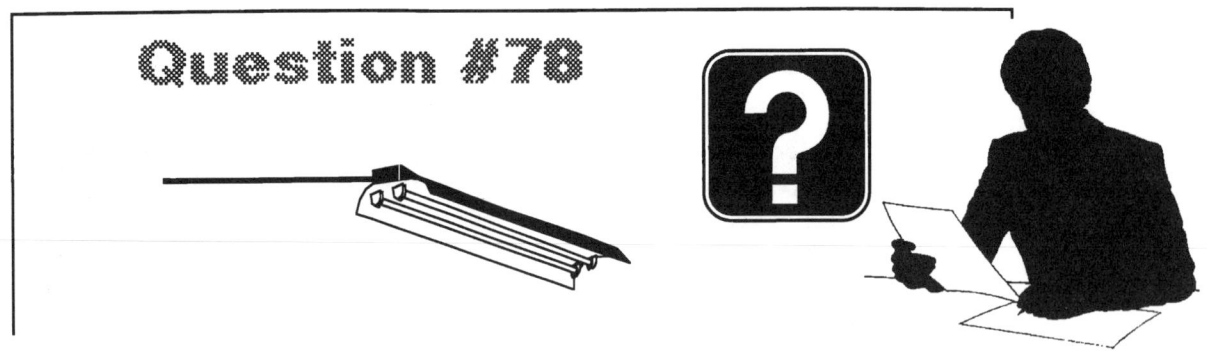

Question #78

The source is 3ø 208/120v. The load is three banks of fluorescent lights connected line to neutral. The three current carrying conductors and neutral are all in the same conduit. If the line current is 50 amps, what size THWN conductors would be required.

(a) 8   (b) 6   (c) 4   (d) 3

Show your work with answer choice:

Question #79

A panel in a dwelling has the following loads:

2 - small appliances	4 - 1920w loads
1 - laundry	1 - 12kw range
1 - 1.2kw dishwasher	1 - 4kw dryer

Which panel is best balanced?

(a)
A	N	B
1500	5280	1200
1920	3720	1920
8000		8000

(b)
A	B
1500	1500
1500	1920
1920	1920
1920	2000
2000	4000
4000	1200

(c)
A	B
1500	3000
1920	1920
1920	1920
1200	2000
2000	4000
4000	

(d)
A	B
3000	1500
1920	1920
1920	1920
2500	2500
4000	4000
1200	

Show your work with answer choice:

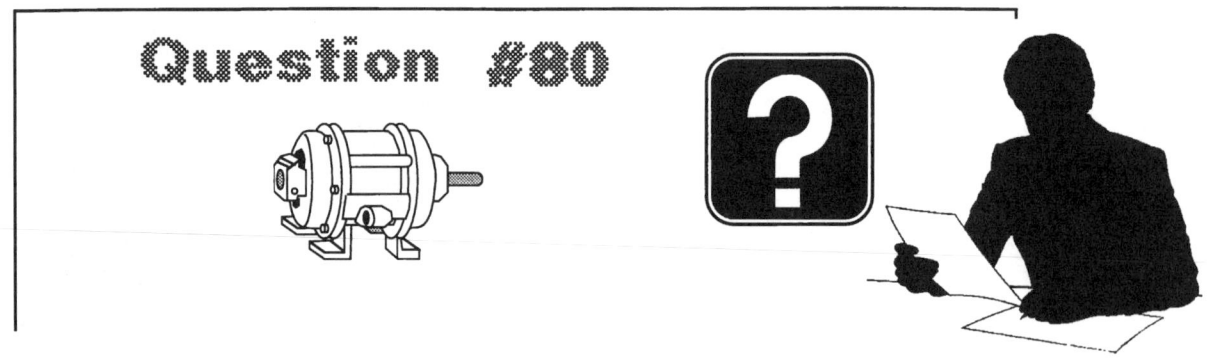

Question #80

A 200 hp, 2300 volt, three-phase motor is delivering 100 hp, with full load amps of 49. How many kvars are needed to bring the power factor up to .95 lagging?

(a) 74 (b) 195 (c) 180 (d) 156

Show your work with answer choice:

CALCULATIONS II

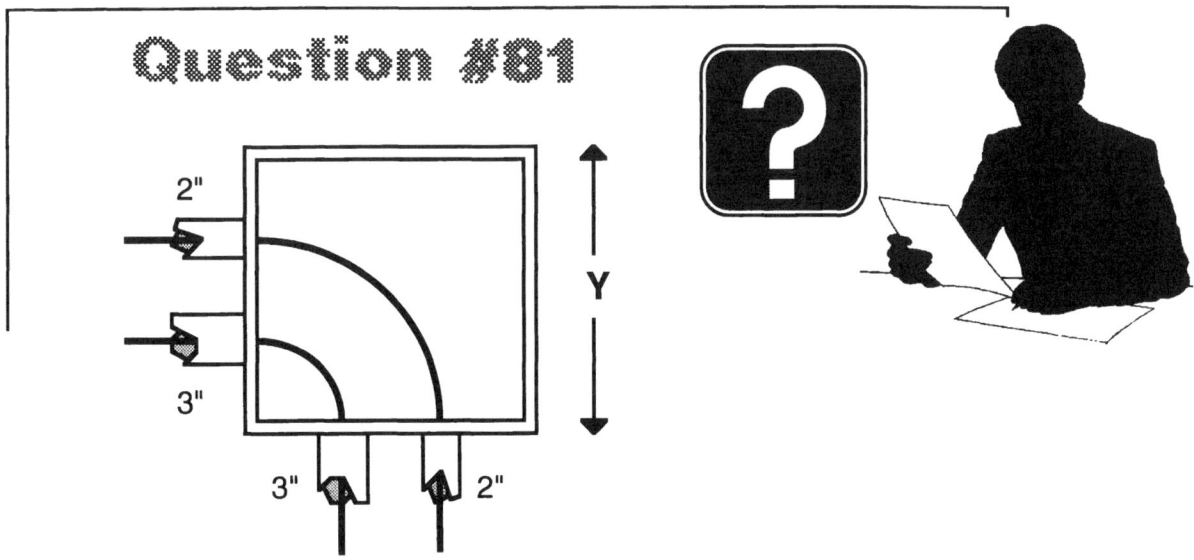

In the following drawing of a pull box, the dimension "Y" should not be less than _____ inches.

(a) 15   (b) 18   (c) 20   (d) 24

Show your work with answer choice:

Question #82

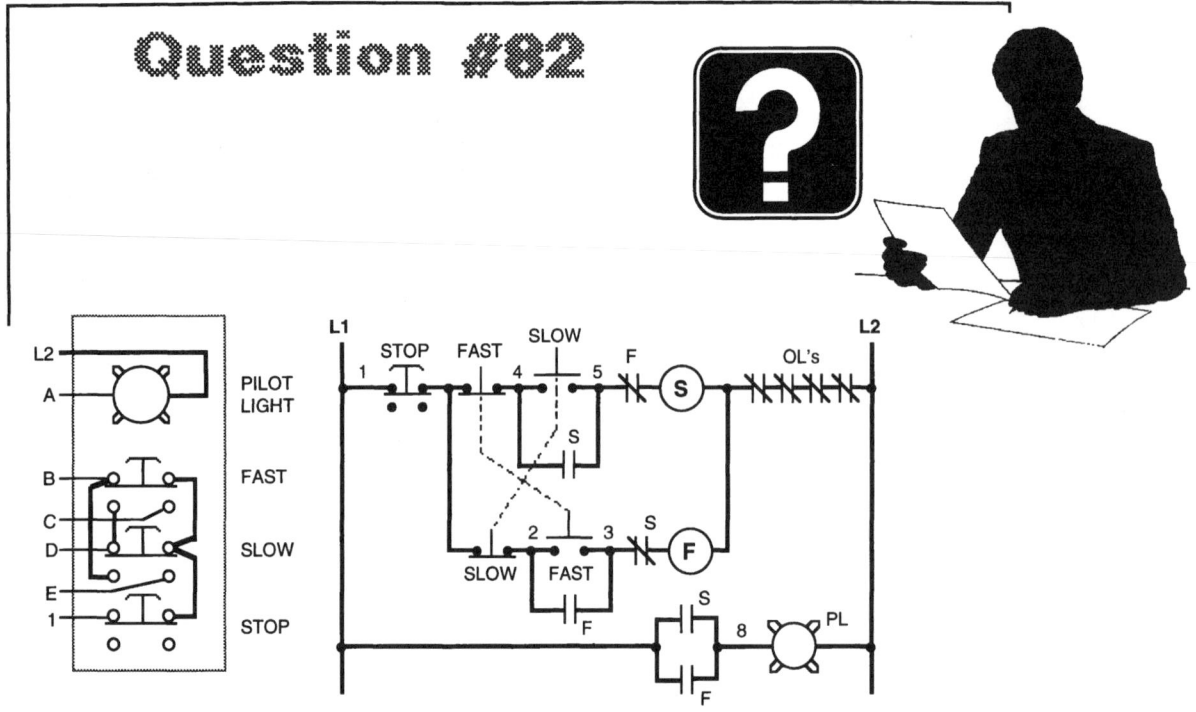

"**A**" is actually point _____ on the control diagram.

(a) **L1**   (b) **L2**   (c) **8**   (d) **6**

Show your work with answer choice:

Question #83

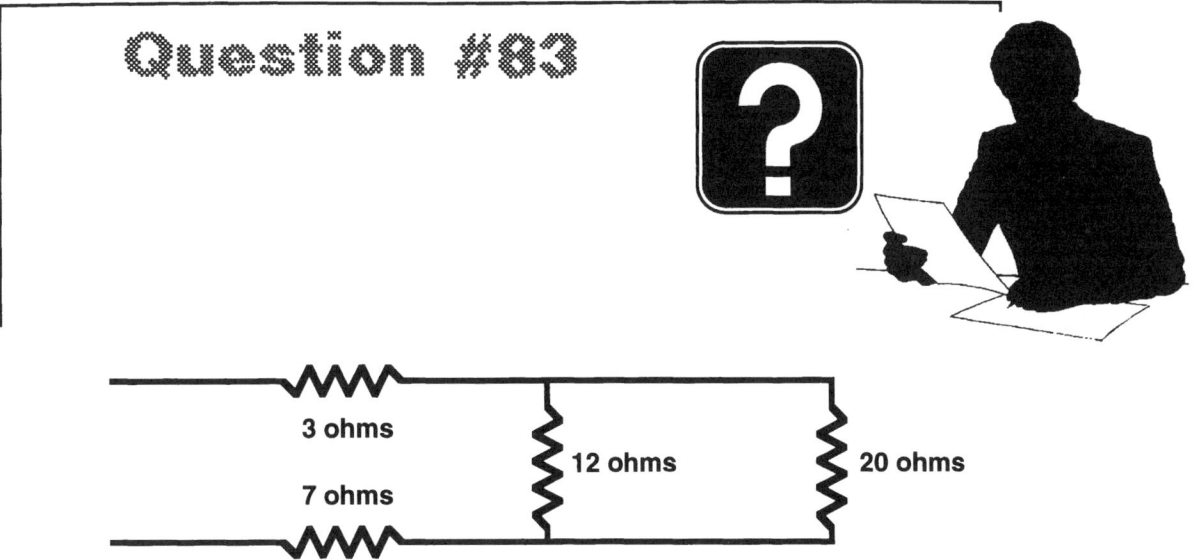

What is the total resistance in the series-parallel circuit?

(a) 42Ω (b) 4.28Ω (c) 17.5Ω (d) none of these

Show your work with answer choice:

Question #84

What is the minimum amperage required for the following boat berth receptacles?

15 - 15 amp 120v locations
12 - 30 amp 120v locations
 8 - 50 amp 120v locations

_____ **amps.**

Show your work

CALCULATIONS II

A spot welder supplied by a 60-hertz system makes three hundred 20-cycle welds per hour. What is the duty cycle percent for this welder?

(a) 3.6%   (b) 3%   (c) 2.8%   (d) none of these

Show your work with answer choice:

Question #86

A 7200/240 volt transformer has 1800 turns in its primary winding. The number of turns in the secondary is ____.

(a) 45   (b) 60   (c) 956   (d) 1800

Show your work with answer choice:

CALCULATIONS II

What is the minimum neutral in amps for a 2000 sq.ft. house with the following:

(1) 120/240 volt service
(2) no electrical appliances
(3) assume totally balanced load

(a) 16.4 amps   (b) 23.4 amps   (c) 25 amps   (d) 43.8 amps

Question #88

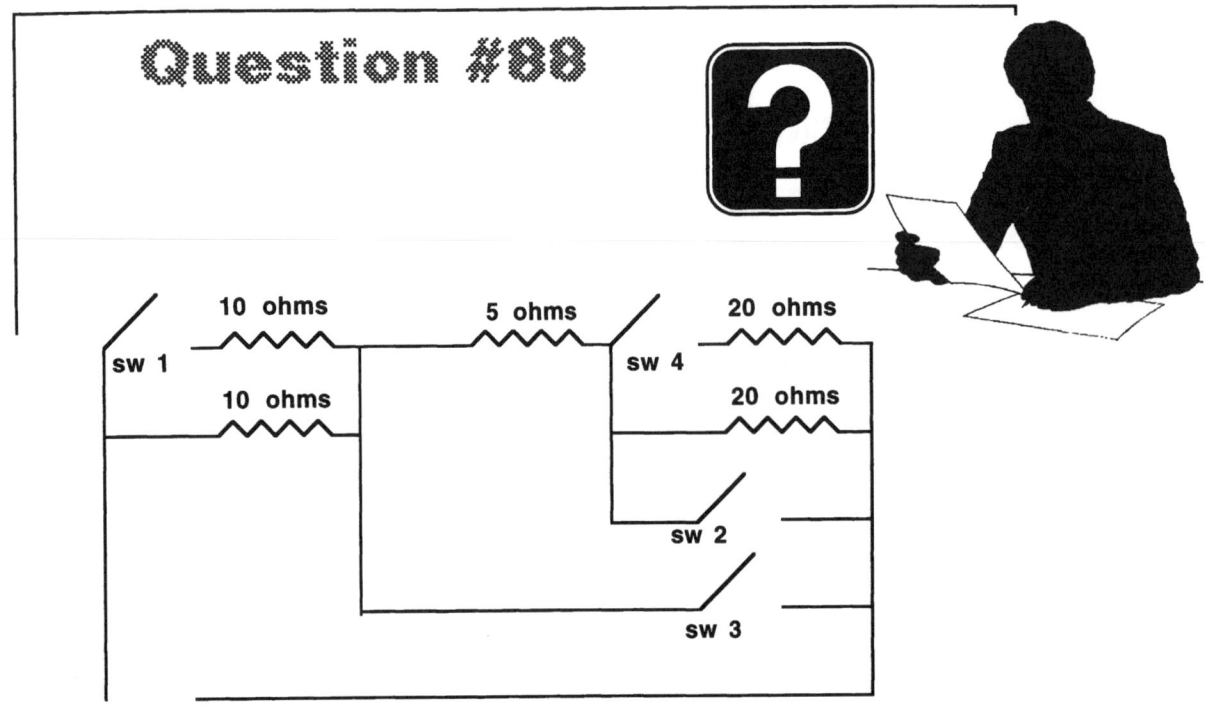

With only switch 4 closed and a line voltage of 225 volts, the drop across one of the 10 ohm resistors is ____ volts.

(a) 225   (b) 90   (c) 64.3   (d) 56.3

Show your work with answer choice:

CALCULATIONS II

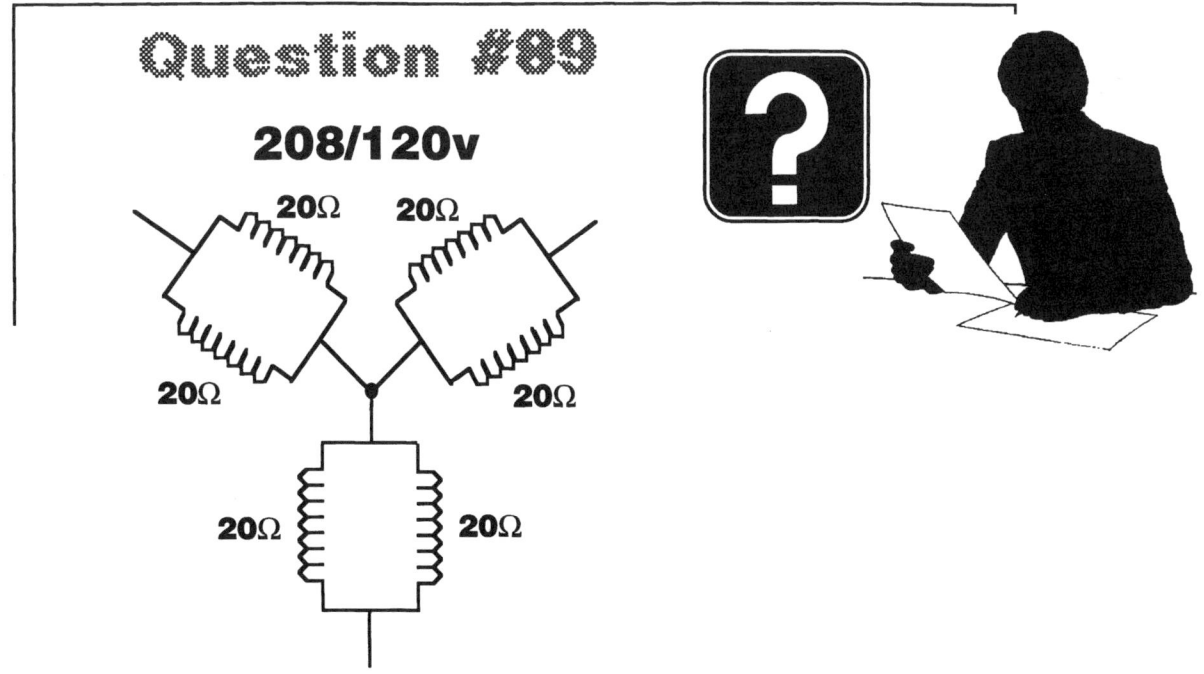

Question #89

208/120v

What is the approximate kva of this transformer?

(a) 1.08   (b) 2.16   (c) 4.32   (c) 8.64

Show your work with answer choice:

CALCULATIONS II

Question #90

A restaurant with a 208/120 volt 4 wire 3 phase service has the following:

1- 25 kw fryer
1 - 2 kw toaster
1 - 6 kw booster heater
1 - 24 kw oven
1 - 10 kw dishwasher
1- 2 kw bun warmer
lights at 125% and are rated 4 kva
62 general use receptacles at 180 va
The demand on the service for this equipment would be ____ amps.

(a) 125 - 150   (b) 151 - 175   (c) 176 - 200   (d) 201 - 250

Show your work with answer choice:

CALCULATIONS II

A 60 unit apartment complex has a 1200va dishwasher rated at 120 volts 1 ø in each unit. What is the demand in amperes required on the 240/120v 1 ø service for these dishwashers?

(a) 225 amps   (b) 300 amps   (c) 450 amps   (d) 600 amps

Show your work with answer choice:

CALCULATIONS II

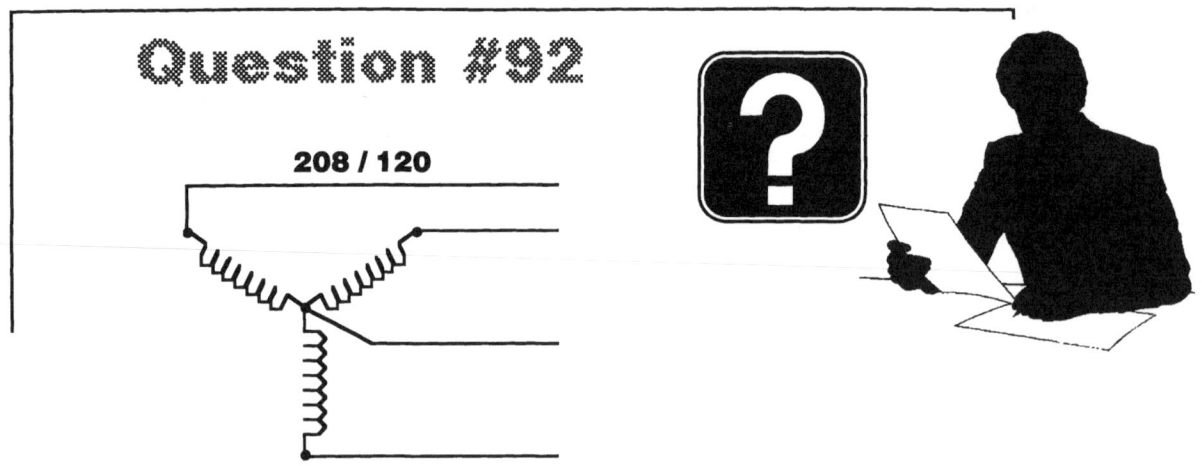

Question #92

208 / 120

What is the current flow in amps for the neutral of a wye three-phase system given the three individual line currents flowing are: L1 = 20 amps, L2 = 35 amps, and L3 = 40 amps.

(a) 17.32 amps (b) 18.02 amps (c) 20 amps (d) 30 amps

Show your work with answer choice:

CALCULATIONS II

Question #93

What is the branch circuit demand for a 8 kw wall-mounted household oven 240/120 volts?

(a) 8 kw   (b) 6.4 kw   (c) 5.6 kw   (d) 4.48 kw

Show your work with answer choice:

Question #94

The voltage drop on two #12 solid uncoated THW conductors, 150 feet long, connecting a 9.8 amp load to a 115 volt source would be ____ volts.

(a) 3.45   (b) 5.75   (c) 5.6742   (d) none of these

Show your work with answer choice:

CALCULATIONS II

A store has 30 feet of show window. What is the amperage required on the service for the show windows? 120/240v 1ø service.

(a) 25 amps   (b) 50 amps   (c) 31 amps   (d) 62.5 amps

Question #96

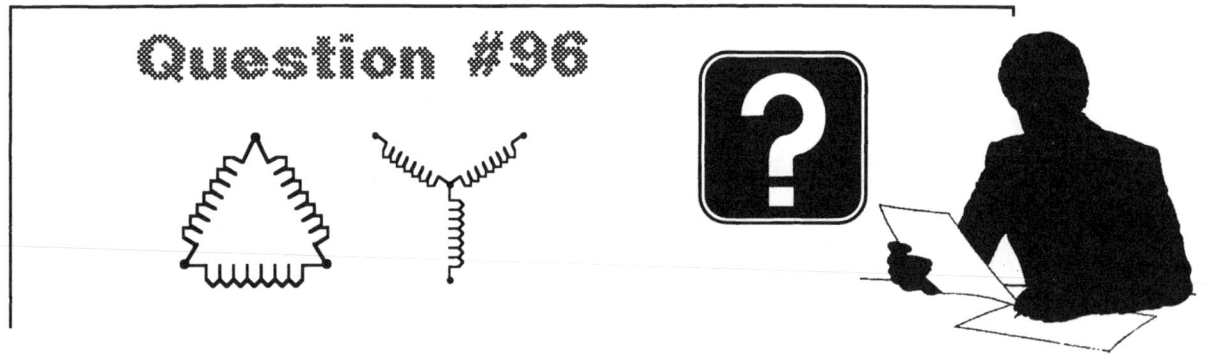

If the source is 3ø, 480/208/120v and the delta-wye transformer has a total load of 50 kva, 3 ø, what is the primary line current?

(a) 35 amps   (b) 60 amps   (c) 104 amps   (d) 123 amps

Show your work with answer choice:

CALCULATIONS II

In a motel, the following spaces exist:

(1) 100 guest rooms that measure 12' x 20' each
(2) total of 2000 sq.ft. hallways
(3) beauty shop 20' x 20'
(4) office 20' x 60'

The minimum lighting demand would be approximately ____.

(a) 55,600 va   (b) 28,800 va   (c) 28,300 va   (d) 30,400 va

Show your work with answer choice:

Question #98

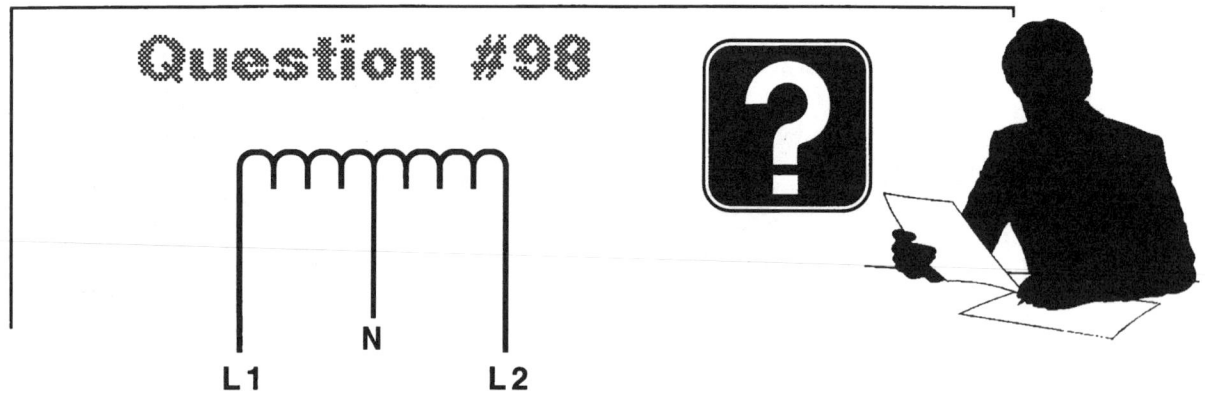

If the total load on a center tapped 240/120v transformer is 100 kw, and the 120v loads were balanced and the 240v load on this phase of the transformer was 20 kw, the maximum neutral current, under any condition, would be approximately ____ amps.

(a) 0   (b) 192   (c) 333   (d) 667

Show your work with answer choice:

CALCULATIONS II

Question #99

In an office building what is the maximum number of duplex receptacles that can be installed on a 20 amp, 120v branch circuit?

(a) 6   (b) 13   (c) 14   (d) unlimited

Show your work with answer choice:

CALCULATIONS II

Question #100

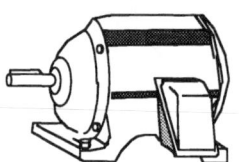

What size dual-element fuse does the Code require for a 2hp, 208 volt, single-phase motor?

(a) 20 amps   (b) 30 amps   (c) 35 amps   (d) 40 amps

Show your work with answer choice:

Question #101

Use the general method of calculation and determine the demand on the service for the following school building. 208/120v single-phase service.

20,000 sq.ft. of classroom
4,000 sq.ft. auditorium
2,000 sq.ft. cafeteria
10 kw outside lighting 120v
5 kw stage lighting 120v
200 receptacles 120v

Cafeteria equipment:
2 - 14 kw ranges 208v
2 - 6 kw ovens 208v
3 - 4 kw fryers 208v
1 - 12 kw water heater 208v
1 - 3 kw dishwasher 208v
1 - 6 kw booster heater 208v
2 - 2 kw toasters 120v
2 - 1/2 hp hood fans 120v
2 - 3/4 hp grill vent fans 120v
4 - 10 hp A/C units 208v
40 kw electric heat 208v

Show your work

ANSWERS - CALCULATIONS II

ANSWERS - CALCULATIONS II

Question #1 Solution:

Use the Ohms Law power wheel which shows I = W/E

With no power factor mentioned W = VA

Since it is three-phase the 1.732 is used

To solve amperes: I = VA/E x 1.732

k = kilo (1000) to convert to va multiply the kva by 1000

167 kva x 1000 = 167,000va

167,000va/208v x 1.732 = 463.5 amperes

The next step is to find the multiplier. Divide 100 (a constant) by the % of impedance. 100/2% = 50. The multiplier is 50.

Multiply the current of 464 amps x 50 = **23,200** amperes is the available fault current.

KEY:

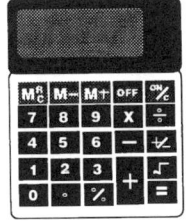

How to use your calculator with formulas must be understood first. 167,000va/ 208v x 1.732 is written correctly. The 167,000va must be divided by (208 x 1.732) which equals 360.256. 167,000/360.256 = 463.55924.
Or - the way I use my calculator is to press 167,000 - press the *divide* button - press 208 - press the *divide* button - press 1.732 - press equals = 463.55924

I use my calculator this way rather than the first step as it has been proven the least amount of movements with the hands the less chance of an error. It's not for speed, it's for accuracy.

Question #2 Solution:

Turn to Table 430.150 to find the full load current of the 25 hp, 208v, 3ø motor which is 74.8 amperes.

Since the application of the motor is used on a freight elevator and is a 15 minute rated motor the next step is to turn to Table 430.22(E), Exception. This Table lists a 15 minute rated motor at 85% of the nameplate current rating.

74.8 amps x 85% = 63.58 required ampacity.

Turn to Table 310.16 and the first column is for the TW insulation of 60° C. A **#4 TW** with an ampacity of 70 is the minimum size conductor permitted.

KEY:

Table 430.22(E), Note.

Any motor application shall be considered as continuous duty unless the nature of the apparatus it drives is such that the motor will not operate continuously with load under any condition of use.

ANSWERS - CALCULATIONS II

Question #3 Solution:

Turn to Table 310.16 to the second column which is the 75°C ampacity.

For a 75 amp load it would require a **#4** with an ampacity of 85 amperes. This is the smallest size THHN conductor permitted connected to 75°C devices.

This is a very good question and should be asked more often on exams as this is a very often violated section of the Code in the everyday field installation of circuits using the THHN wire.

KEY:

Code section 110.14(C) states for loads 100 amps and less 60°C conductors and over 100 amps 75°C conductors. There is no 90°C rated devices. The 90°C conductors can be used, *but they can only be loaded to the rating of the devices either 60°C or 75°C.*

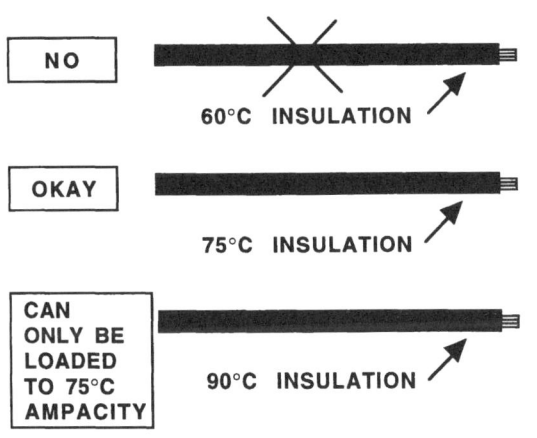
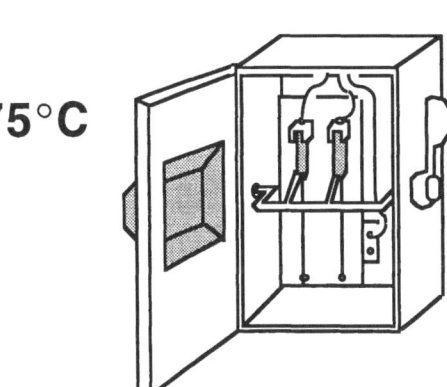

ANSWERS - CALCULATIONS II

Question #4 Solution:

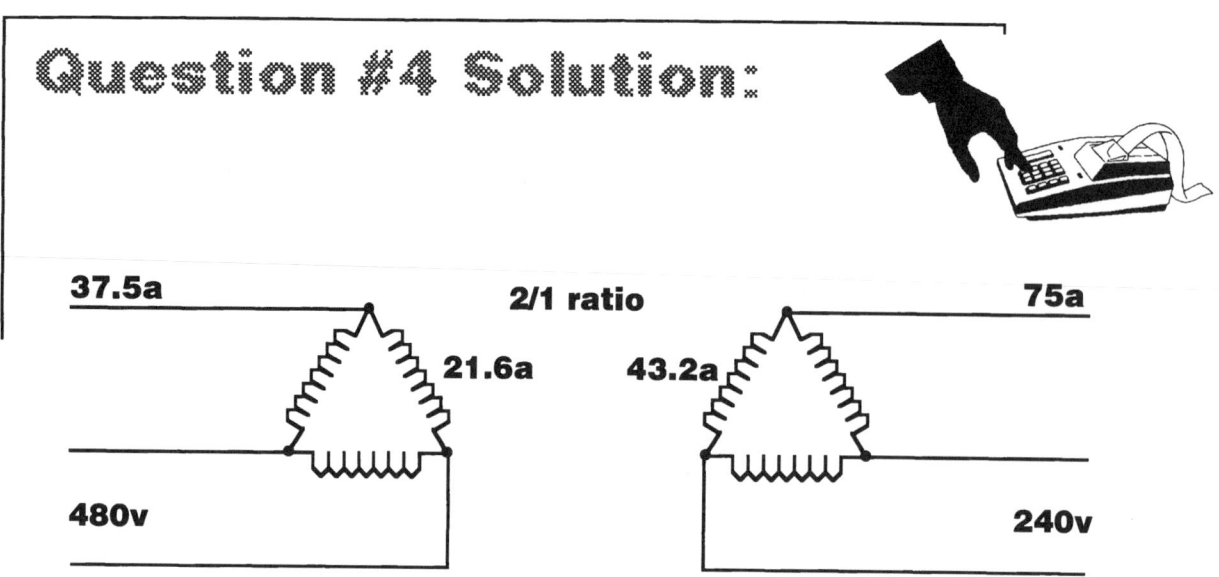

The ratio of transformation is 480v/240v = 2 to 1 ratio. 75a/2 = **37.5** primary amps.

KEY:

Before ever taking a master electrical exam be sure to work the book "Transformer Exam Calculations". This book clearly explains these difficult calculations and has easy to use formats as shown below.

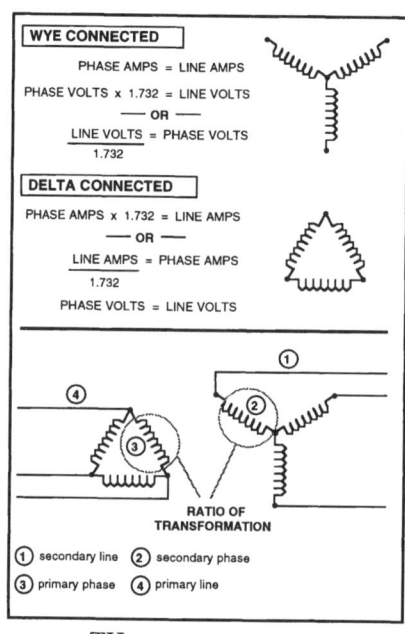

ANSWERS - CALCULATIONS II

Question #5 Solution:

1st step: Find power factor   PF = W/VA = 1630w/2400va = 68%

2nd step find inductive vars:

vars = $\sqrt{va^2 - w^2}$ = 2400va x 2400va = 5,760,000va   1630w x 1630w = 2,656,900w now subtract the watts from the volt amps = 5,760,000va - 2,656,900w = 3,103,100 now press the square root button on your calculator $\sqrt{}$ = 1762 vars

3rd step find the va required to raise power factor to 95%: 1630w/.95 = 1716 va required.

4th step:

vars = $\sqrt{va^2 - w^2}$ = 1716va x 1716va = 2,944,656va   1630w x 1630w = 2,656,900w now subtract the watts from the volt amps = 2,944,656va - 2,656,900w = 287,756 now press the square root button on your calculator $\sqrt{}$ = 536 vars

5th step: 1762 - 536 = **1226 vars** needed to raise the PF to 95%.

KEY:

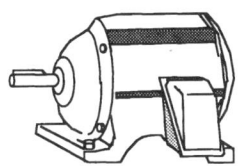

This question can be solved with a standard 8 digit calculator.

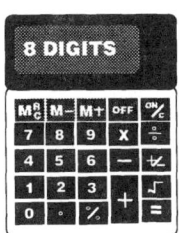

vars = $\sqrt{va^2 - w^2}$

5th step the present inductive vars are 1762, to find the needed vars to produce a reactive power of 536 vars subtract the amount of needed vars from the present amount of vars = 1762 vars - 536 vars = *1226 vars or 1.226 kvars needed to raise to 95% power factor.*

This is a calculation you need to practice a few times.

ANSWERS - CALCULATIONS II

Question #6 Solution:

Solution: $I_n = \sqrt{I^2A + I^2B + I^2C - (IA\ IB) - (IB\ IC) - (IC\ IA)}$

The formula at first looks very difficult, but really it's not.

Everything under the square root sign must be done first, which means:
current in A squared = 60 x 60 = 3600
+ current in B squared = 70 x 70 = 4900
+ current in C squared = 80 x 80 = 6400
                                            14900 total   (call this total "X")

Now the right side of the formula shows:
current in A x current in B = 60 x 70 = 4200
current in B x current in C = 70 x 80 = 5600
current in C x current in A = 80 x 60 = 4800
                                            14600   (call this total "Y")

Now subtract total "Y" from total "X" = 14900 total X
                                        — 14600 total Y
                                           300

Now extract the square root by pressing the √ button on your calculator.
The answer **17.32** amps is the unbalanced current flowing in the neutral.
17.320508 is the square root of 300. 17.320508 x 17.320508 = 300.

KEY: NEUTRAL CURRENT IN A WYE

$$I_n = \sqrt{I^2A + I^2B + I^2C - (IA\ IB) - (IB\ IC) - (IC\ IA)}$$

The formula looks difficult at first, but after working a few calculations it's actually very simple. The first one is the toughest!

ANSWERS - CALCULATIONS II

Question #7 Solution:

The first step is to balance the 30 ranges among the three phases.

A phase = 10 ranges   B phase = 10 ranges   C phase = 10 ranges

The 208v range would connect between two phases. Always take the largest number of ranges on any one phase times 2 (the connection between two phases).

10 ranges x 2 = 20 ranges. Instead of 30 ranges we now count only 20 ranges, this is the advantage of using a three-phase system instead of single-phase.

Turn to **Table 220.19** for Household cooking equipment demand and Column C shows for 20 appliances the demand would be **35 kw.**

KEY:

For **10** ranges, find the best balance per phase:

A PHASE   B PHASE   C PHASE
4 ranges 3 ranges 3 ranges

```
       SINGLE-PHASE                    THREE-PHASE
         SYSTEM                          SYSTEM

    L1           L2              A        B        C


       CONNECTIONS                  CONNECTIONS
        L1 — L2 = 10                A — B = FOUR
                                    B — C = THREE
                                    A — C = THREE
```

THE MOST CONNECTIONS
| **FOUR x 2 PHASES = 8 APPLIANCES** |

Take the largest number of ranges on one phase (4 on A PHASE) times the number of connections (two). 4 ranges x 2 = **8 appliances**. This is the advantage of a 3-phase over a single-phase, instead of calculating 10 appliance loads, you now have **8** appliance loads.

ANSWERS - CALCULATIONS II

Question #8 Solution:

Table 220.3(A)   office building

Lighting in an office is considered a continuous load and 210.20(A) states a circuit can only be loaded 80%.

40,000 sq.ft. x 3.5va   =   140,000va = 73 circuits for lighting
20a x 120v = 2400va x 80%   =   1920va

*An additional 1va from Table 220.3(A) is required for unknown receptacles.

40,000 sq.ft. x 1va   =   40,000va   = 17 circuits for receptacles
20a x 120v   =   2400va

2400va can be used for each 20 amp receptacle circuit since receptacles are not considered continuous.

A total of **90 circuits** are required.

KEY:

Lighting is considered a continuous load and receptacles are NOT.

The question is asking for the number of circuits, not the service demand.

For the service demand : 40,000 sq.ft. x 3.5va x 125% = 175,000va + 40,000 va for unknown receptacles = 215,000 va. But the question is asking for the number of circuits required so the 4.5va is split, 3.5va for continuous lighting circuits and 1va for the noncontinuous receptacle circuits.

Question #9 Solution:

Code section 550.31(1) requires 16,000 va (16 kva) for each mobile home.

16 kva x 25 sites = 400 kva x 24% Table 550.31 = 96 kva

KEY:

Table 550.31 is a demand factor for feeders and service entrance conductors. 25 mobile homes has a 24% demand factor.

ANSWERS - CALCULATIONS II

Question #10 Solution:

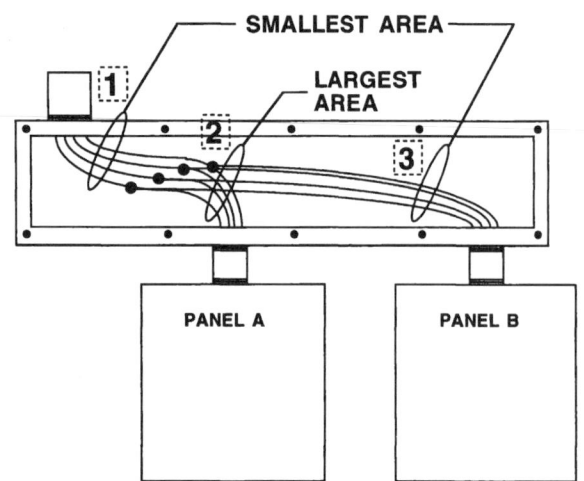

To size the gutter required, the largest area of conductor concentration must be used. Code Table 5 lists the area square inch for conductors with THHN insulation.

The area of concentration at [1] would be 4 - #300 THHN conductors. Table 5 lists a #300 at .4608 sq.in. x 4 conductors = 1.8432 cross sectional area.

The area of concentration at [2] would be 8 - #3/0 THHN conductors. Table 5 lists a #3/0 at .2679 sq.in. x 8 conductors = 2.1432 cross sectional area.

The area of concentration at [3] would be 4 - #3/0 THHN conductors. Table 5 lists a #3/0 at .2679 sq.in. x 4 conductors = 1.0716 cross sectional area.

The largest area of conductor concentration in the gutter would be at point [2] of 2.1432 cross sectional area.

The Code states the most the gutter can be filled is 20% (the reciprocal is 5) multiply the largest conductor area of 2.1432 x 5 (reciprocal of 20%) = 10.716 cross sectional area required.

This requires a gutter of an area of at least 10.716 square inches. A *4" x 4"* gutter would provide an area of 16 sq.in. Table 373.6(A) must also be checked for bending space.

Code Table 312.6(A) lists the minimum wire-bending space and minimum width of gutters. Since the service enters the top of the gutter rather than the end, the table requires a minimum bending depth of 5" for a single #300 conductor. **This requires a 6" x 6" minimum gutter.**

KEY:

If the service conductors entered the end instead of the top, Table 312.6(A) would be applied to the bending space of the #3/0 conductor going out of the bottom of the gutter. The table requires a bending space of only 4" for a #3/0 conductor.

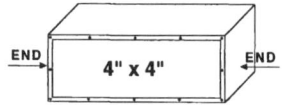

A 4" x 4" gutter would be suitable if the service conductors entered the end of the gutter instead of the top.

Question #11 Solution:

$$\frac{750{,}000\,va}{25{,}000\,sq.ft.} = 30\,va\text{ per sq.ft.}$$

Table 220.34:   1st 3va @ 100% = 3va
Next 17va @ 75% = 12.75va
Next 10va @ 25% = 2.5va
18.25 va per sq.ft.

18.25va per sq.ft. x 25,000 sq.ft. = **456,250va**

KEY:

The first step is to divide the total load by the total square footage and then apply the demand factor to this number.

Optional method for a school is Code section 220.34. Table 220.34 permits a demand to be applied above 3 va.

ANSWERS - CALCULATIONS II

Question #12 Solution:

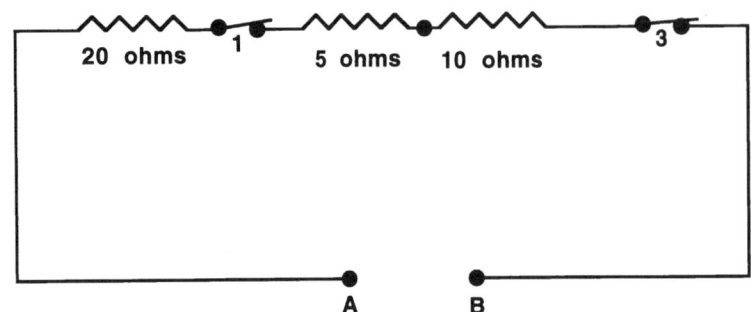

With switches (1) and (3) closed the series circuit has a total resistance of **35 ohms.**

KEY:

In a series circuit resistances add together for a total.

Question #13 Solution:

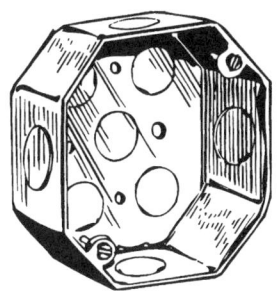

#12 wires = 4 x 2.25 cubic inches (Table 314.16B) =   9 cubic inches
#14 wires = 3 x 2 cubic inches (Table 314.16B)   =   6 cubic inches
                                                                                    15 cubic inches

Table 314.16(A) = **1 1/2" octagon box** for 15 cubic inches.

KEY:

Both Table 314.16(A) and 314.16(B) are used for this calculation.

Question #14 Solution:

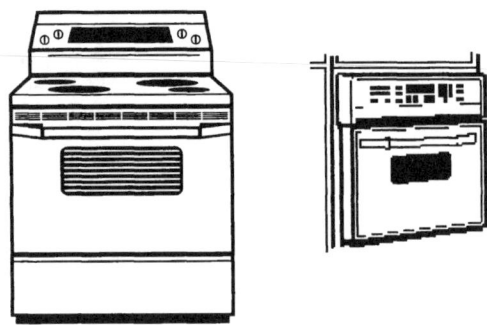

Table 220.19 Household cooking equipment demands. Column C for two appliances = **11 kw.**

KEY:

You are always calculating to the lowest value the Code allows with a demand. If you used Column C for the range at 8 kw and Column B for the oven at 3.2 kw (4kw x 80%) the total is 11.2 kw.

Now with two 12 kw ranges simply go to Column C and the demand for two appliances is 11 kw, which is less than 11.2 kw.

ANSWERS - CALCULATIONS II

Question #15 Solution:

46 lights x 250 watts = 11,500 watts total.

The Code in section 384-16d states a circuit cannot be loaded over 80% for continuous loading.

20 amp circuit x 120 volts = 2400 watts x 80% = 1920 watts per circuit.

11,500w/1920w = 5.98 or 6 circuits, but when dividing the lights between the three phases A = 16 lights, B = 15 lights, C = 15 lights. This would violate the 80% loading for continuous. Each light 250w/120v = 2.08 amps x 8 lights = 16.64 amps. Phase A would require 3 circuits, Phase B would require 2 circuits, Phase C would require 2 circuits for a total of **7 circuits**.

KEY:

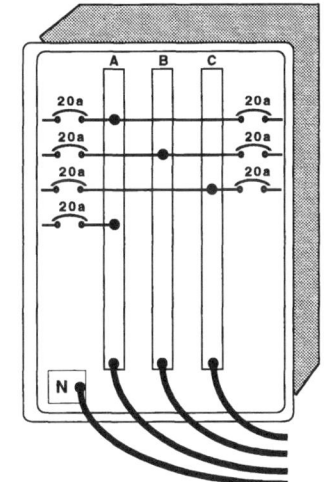

THREE-PHASE 4-WIRE

The 7 circuits are connected to the three phases.

   Phase A has 3 - 20 amp circuits
   Phase B has 2 - 20 amp circuits
   Phase C has 2 - 20 amp circuits

Each light is 250 watts/120v = 2.08 amps

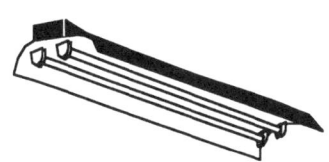

The 46 lights total must be divided among the 7 circuits so that no circuit is loaded to more than 16 amps continuously (20a x 80%).

Phase A Circuit 1 has 6 lights x 2.08 amps = 12.48 amps total
Phase A Circuit 2 has 6 lights x 2.08 amps = 12.48 amps total
Phase A Circuit 3 has 6 lights x 2.08 amps = 12.48 amps total

Phase B Circuit 1 has 7 lights x 2.08 amps = 14.56 amps total
Phase B Circuit 2 has 7 lights x 2.08 amps = 14.56 amps total

Phase C Circuit 1 has 7 lights x 2.08 amps = 14.56 amps total
Phase C Circuit 2 has 7 lights x 2.08 amps = 14.56 amps total

Question #16 Solution:

#2/0 THW = .2624 x 4 =   1.0496
#6   THW = .0726 x 6 =   .4356
#12 THW = .026 x 20 =   .52

2.0052 square inches

KEY:

Table 5 lists the area of square inch for insulated conductors.

A wireway is a sheet-metal trough used instead of conduit to run large wires a considerable distance to several locations. The cover may be hinged or removable.

A gutter can only extend 30 feet from the equipment. A wireway has no limit on the distance it is run. Wireways are available in lengths of 1, 2, 3, 4, 5, and 10 feet, and in cross sections of 2 1/2" x 2 1/2" up to 8" x 8".

The conductor fill and sizing follow the same 20% and 75% fill area rules for a gutter.

ANSWERS - CALCULATIONS II

Question #17 Solution:

Section 440.22(A) requires a 175% of the compressor rated-load current.
42a x 175% = 73.5a
Section 240.6 lists the next higher standard size at **80 amps**.

KEY:

•Section 440.2. Definition of Hermetic refrigerant motor-compressor: A combination consisting of a compressor and motor, *both* of which are enclosed in the same housing with no external shaft or shaft seals; *the motor operating in the refrigerant.*

The rules of Article 440 apply to hermetically sealed motor-compressors. Whereas the rules of Article 430 apply where the compressor is belt driven.

Most equipment used today is the hermetic type which includes drinking fountains, household refrigerators, freezers, room air conditioners, etc.

Hermetic systems have the motor operating in the refrigerant whereas ordinary motors are cooled by the surrounding air. Hermetically sealed motors can safely consume more current and deliver more horsepower than an ordinary motor of the same size cooled by the ambient air.

While the hermetic motor-compressor is running, the refrigerant is continuously entering the case in a gaseous state that is a cooler temperature than the surrounding air.

After the hermetic motor-compressor sits idle for a period of time the refrigerant reaches a temperature of that of the ambient and provides poor cooling for a short period of time on start up. Consequently the motor heats up faster than an ordinary motor. This is the reason for the different rules of Articles 430 and 440.

Question #18 Solution:

Table 5 lists the area of square inch for insulated conductors.

#500 kcmil XHHW	=	.6984 x 3	=	2.0952 sq.in.
#8 THW	=	.0556 x 3	=	.1668 sq.in.
#14 THHN	=	.0097 x 6	=	.0582 sq.in.

2.3202 sq.in. required

Table 4 requires a **3"** minimum.

KEY:

Table 4 is for 12 different types of raceways. Make sure you are in the heading "Rigid Metal Conduit". The column to select is "Over 2 wires 40%".

A 2 1/2" conduit would allow 1.946 area square inch. The area required is 2.3202, so a 2 1/2" is not large enough.

A 3" conduit @ 40% fill allows 3 square inches.

Question #19 Solution:

$$\frac{120 \times 60\text{Hz}}{1800 \text{ rpm}} = \textbf{4 poles}$$

KEY:

The speed and number of poles of an alternator determine the frequency which it generates.

Where F = frequency in cycles per second, rpm = revolutions per minute of rotor, and P = the number of field poles.

Example: What is the frequency of a two-pole alternator running at 1800 rpm?

Solution: $F = \dfrac{P \times rpm}{120} = \dfrac{2 \times 1800}{120} = $ 30 cycles per second

Question #20 Solution:

$$\frac{7{,}500{,}000 \text{ watts}}{12{,}470\text{v} \times 1.732 \times .75} = 463 \text{ amps closest to } \mathbf{500} \text{ amperes}$$

KEY:

I = W/E

7500 kw = 7,500,000 watts.

The system is three-phase so 1.732 (the square root of 3) must be included in the formula along with the .75 power factor.

ANSWERS - CALCULATIONS II

Question #21 Solution:

The circuit with the switch open looks like this:

Start at the end of the circuit by combining:

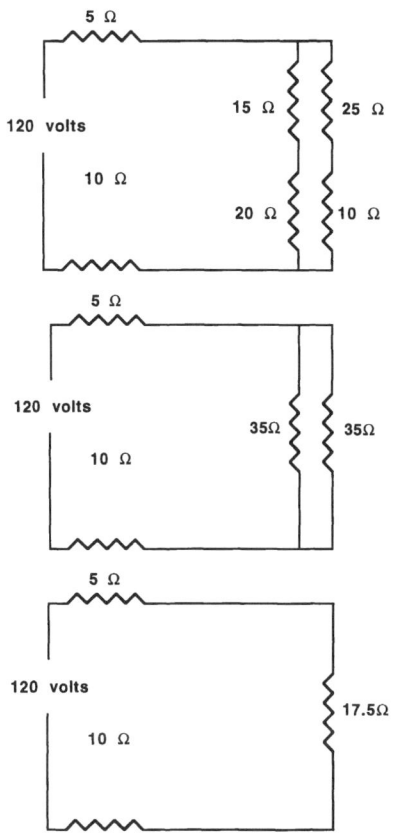

Now you have two 35Ω in parallel which would be 35Ω/2 = 17.5Ω.

Now you add the loads together for the total resistance: 5Ω + 17.5Ω + 10Ω = **32.5** Ω Rtotal.

KEY:

Always start at the end of the circuit and reduce it to it's simplist form.

Question #22 Solution:

Table 250.66 over 1100 kcmil requires a **#3/0 copper.**

KEY:

The 12 1/2% does **not** apply to the sizing of grounding electrode conductors. The 12 1/2% applies to the sizing of bonding jumpers and grounded conductors per 250.102(C) and 250.24(B).

Question #23 Solution:

750 kva transformer, 240v, 3ø, 5 1/2% impedance

$$\frac{750,000va}{240v \times 1.732} = 1804 \text{ amps}$$

1804a x **100** = 180,400/5.5 = **32,800** amperes

KEY:

Formula for short circuit current:

I = $\dfrac{\text{transformer full load amps} \times \mathbf{100}}{\text{transformer \% impedance}}$

Example: 167 kva, 3ø, 208/120v, 2% impedance.

167 kva = 167,000va

$$\frac{167,000va}{208v \times 1.732} = 463.5 \text{ or } 464 \text{ amps}$$

464a x **100** = 46400/2 = 23,200 amperes

Question #24 Solution:

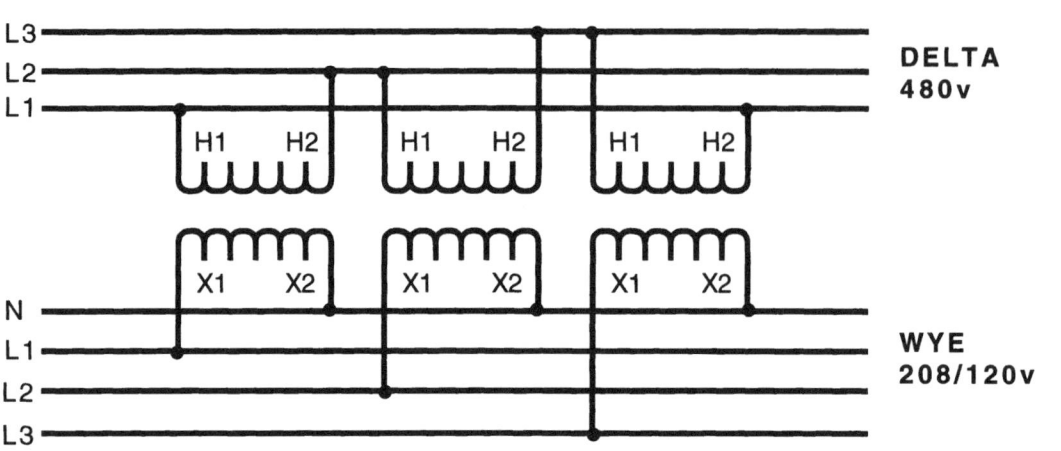

KEY:

The closed delta windings connect together (H1-H2) and the secondary wye, all three windings have one end that connects to the neutral.

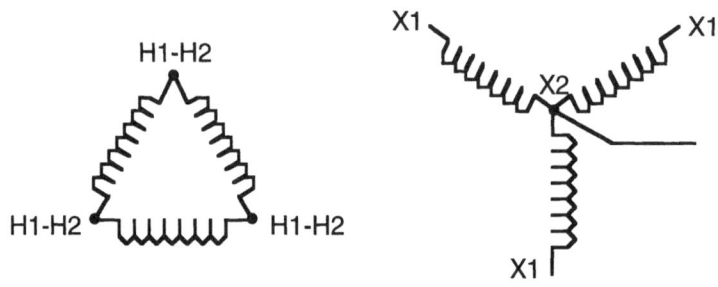

ANSWERS - CALCULATIONS II

Question #25 Solution:

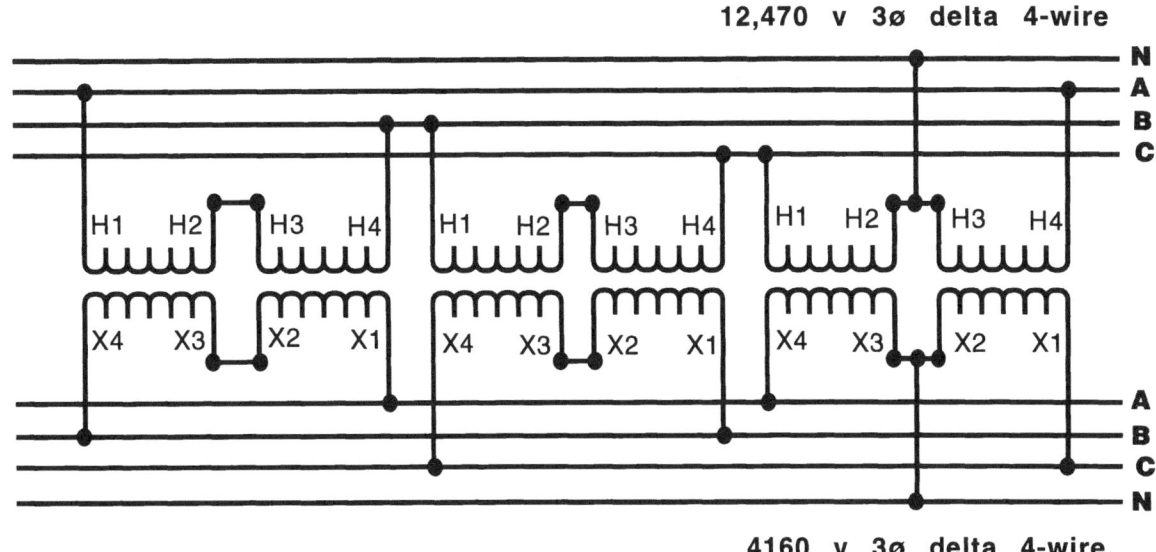

KEY:

The closed delta windings connect H1-H4 to each line with H1-H2 connecting to the neutral on one transformer. On the delta secondary X1-X4 connect to each line and X2-X3 connects to the neutral on one transformer.

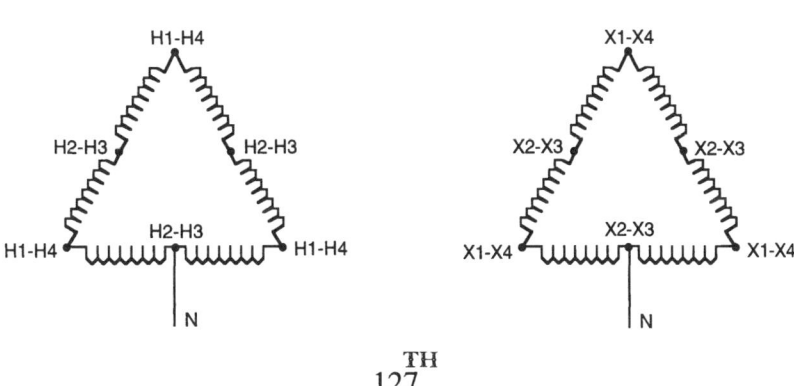

Question #26 Solution:

This is a parallel circuit with unequal resistance. The current flow will be different in each load. To find the current the current meter is reading use I = E/R.

120v/60Ω = **2 amps**

KEY:

To find amps divide the voltage by the resistance.

Question #27 Solution:

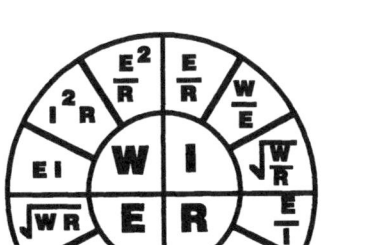

I = E/R

The phase voltage is 120 and the phase resistance is 5 ohms.

120v/5Ω = **24 amps**

KEY:

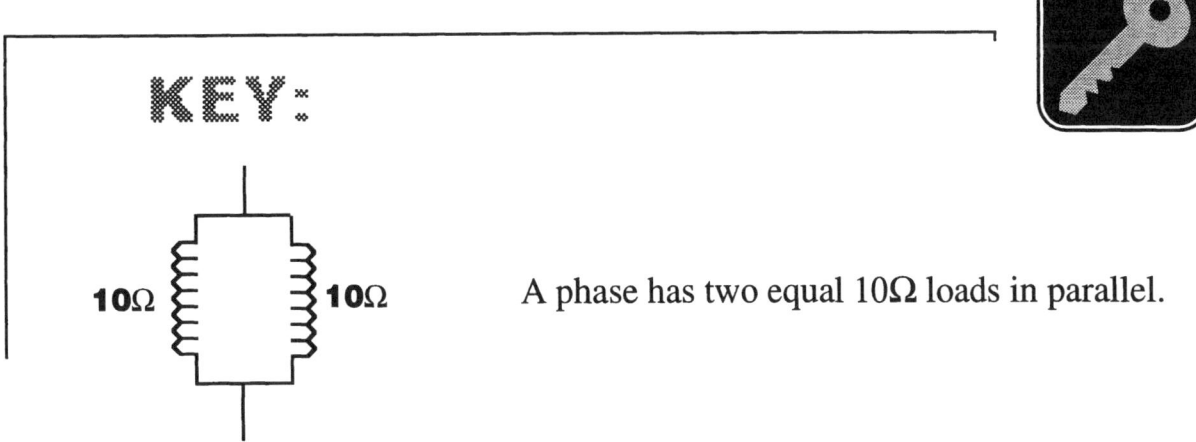

A phase has two equal 10Ω loads in parallel.

To find the phase resistance: 10Ω/2 resistances = 5Ω total phase resistance.

Total resistance in parallel for *equal* resistances = $\dfrac{\text{RESISTANCE of ONE}}{\text{NUMBER of RESISTANCES}}$

ANSWERS - CALCULATIONS II

Question #28 Solution:

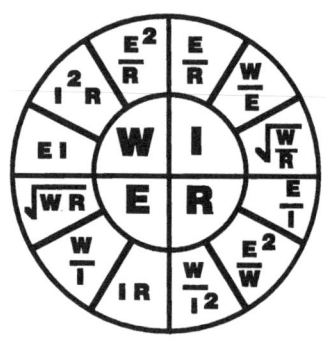

I = W/E = 1500w/120v = **12.5 amps**

#12 wire

20 amp circuit breaker

KEY:

The National Electrical Code section 210.23 allows only 80% on a branch circuit for a single plug and cord connected appliance. A 15a circuit could only be loaded 80% which is 12 amps. The load is 12.5 amps, so a 20 amp circuit is required.

Question #29 Solution:

First find the full-load current of the motor from Table 430.150 3ø.

10.6 amps x 250% (T.430.52 inverse-time breaker) = 26.5 amps

Select a **25 amp** circuit breaker for the branch circuit.

KEY:

The Code permits in 430.52(C1) that you can go up to the **next** higher standard size which would be 30 amps. But 30 amps is NOT one of the answer choices and the Code does NOT permit to go up to a 35 amp as that would be two sizes higher. You are allowed to go up ONLY one size higher. You would drop down to a 25 amp, which is NOT a violation of the Code.

ANSWERS - CALCULATIONS II

Question #30 Solution:

Code section 314.16(B) explains how to count the conductor fill in a box.

2 - black wires
2 - white wires
1 - grounding wires
1 - cable clamps
2 - device
―――――――――
8 - #12 wires

Table 370.16(B)   2.25 cubic inches x 8 #12 wires = **18 cubic inches**

KEY:

Code section 314.16(B1) states you count one for each wire.
314.16(B2) states you count one wire for clamps.
314.16(B4) states you count two for a device.
314.16(B5) states you count one for one *or more* grounding wires.

ANSWERS - CALCULATIONS II

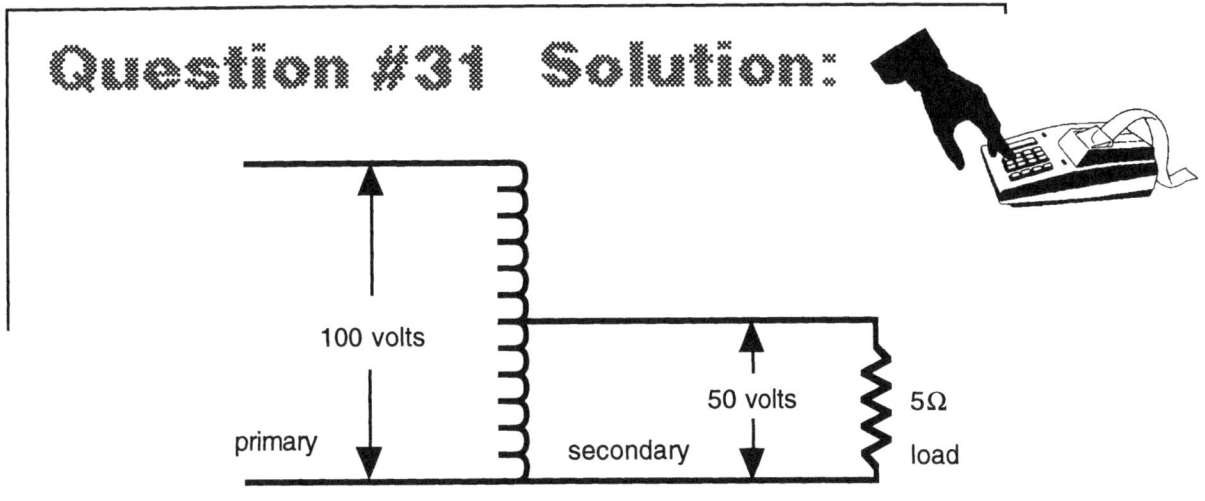

Question #31 Solution:

I = E/R   50v/5Ω = **10 amps**

KEY:

When voltage is stepped **down**, all of the winding is the primary, and part of the winding is the secondary.

When the *ratio of transformation is small* the autotransformer is the *most efficient* and effective. *Less copper* is required for an autotransformer than for a two-winding transformer.

Not economically desirable when the ratio of transformation is *greater* than 2 to 1. Another disadvantage is that the high-voltage and low-voltage systems are electrically connected and could become hazardous to equipment and personnel. The Code forbids the use of autotransformers for lighting and appliance branch circuits unless there is a grounded conductor common to both the primary and secondary circuits.

Question #32 Solution:

First step find I.   I = va/E   9600va/120v = 80 amps

CM = $\dfrac{2 \times 12.9 \times 90' \times 80a}{3.6 \text{ vd}}$ = 51,600 cm required

Table 8 = **#3 conductor**

KEY:

Load "A" is connected single-phase 120 volts. It is a branch circuit which is 3% x 120v = 3.6 voltage drop permitted. Use 12.9 for the approximate K factor.

Question #33 Solution:

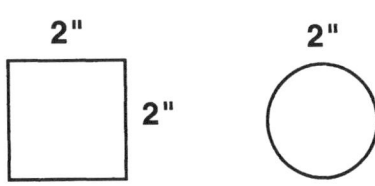 

N.E.C. 366.7(A) - 1000 amps per square inch for a copper bus bar. 2" x 2" = 4 square inches. 4 x 1000a = 4000 amps for a 2" square bus bar. For a *round* bus bar: 4000 amps x *.7854* = 3141.6 amps or **3142 amps.**

KEY:

Bus Bar Square inch area = Width x Thickness

Ampacity = 1000 amps per square inch for copper
700 amps per square inch for aluminum

1 circular mil = .7854 square mils
You can remember this by reading your calculator starting at the top left with "**7**", the number to the right of 7 is "**8**" and below 8 is "**5**" and to the left of 5 is "**4**". By reading this box in a clockwise direction you will remember **7854**.

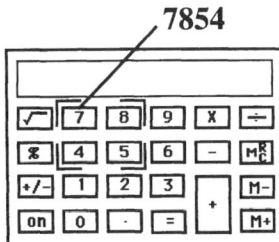

Question #34 Solution:

R = E/I   120v/12a = 10 ohms

$$\frac{\text{Resistance of one}}{\text{Number of resistances}} = \frac{30}{3} = 10\Omega$$

The resistance of R2 would be **30 ohms.**

KEY:

For resistances of equal value in parallel, divide the resistance of one by the number of resistors. The total resistance is 10 Ω, the resistance of each resistor would be 30Ω since they are equal.

Question #35 Solution:

T.430.148 3hp, 208v = 18.7a x 250% = 46.75 = 50a circuit breaker
T.430.150 5hp, 208v = 16.7a x 250% = 41.75 = 45a circuit breaker

Largest branch circuit breaker = 50a + 16.7a F.L.C. = 66.7 amps or **60 amps.**

KEY:

You are NOT permitted to go up to the next higher standard size on a feeder. There is NO exception in 430.62 permitting it. You must drop down to a 60 amp.

ANSWERS - CALCULATIONS II

Question #36 Solution:

2 x 5,000 watt water heaters	=	10kw
4 x 3,000 watt fryers	=	12kw
2 x 6,000 watt ovens	=	12kw
		34kw

34 kw x 65% (8 pieces of equipment T.220.20) = 22.1 kw or 22,100 watts

22,100w/480v x 1.732 = **27 amps**

KEY:

Code section 220.20 is for commercial cooking equipment. Any equipment used in the *operation* of cooking can be counted. A demand from Table 220.20 is permitted.

Question #37 Solution:

I = E/R   = 208v/30Ω = 6.93 phase amps

6.93a x 1.732 = **12 line amps**

KEY:

IL is asking for the current in the line, delta connected. The phase voltage is 208 and the phase load is 30Ω. The phase current is 6.93 amps. To find delta line amps the phase amps must be multiplied by the square root of three (1.732).

Question #38 Solution:

The lowest resistance is when switches **2, 3, and 4 are open.**

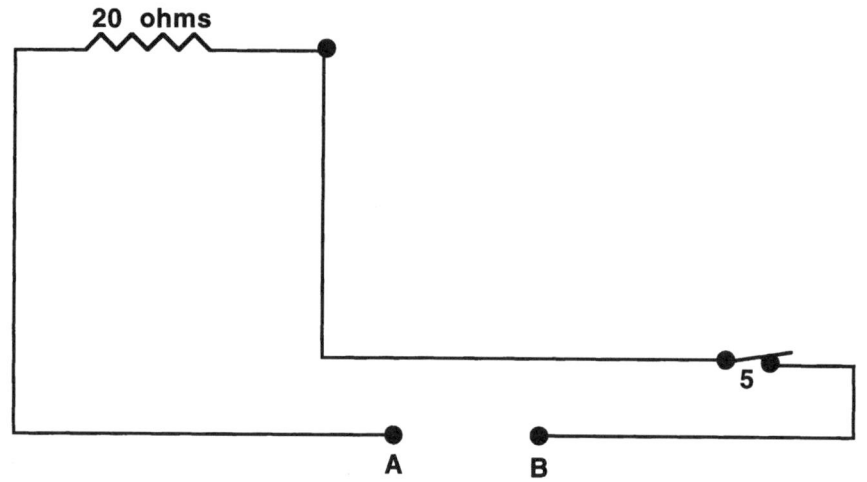

KEY:

When switches 2, 3, and 4 are open, switches 1 and 5 are closed. With switch 5 closed the circuit has a resistance of 20 ohms. Switch 1 is also closed but does not add any resistance to the circuit.

ANSWERS - CALCULATIONS II

Question #39 Solution:

Table 430.150 = 34a F.L.C. x 125% = 42.5a + 30.8 = 73.3 required ampacity

Table 310.16 = **#3 TW**

KEY:

The F.L.C. of the largest motor is increased 25% plus all the other motors connected to the SAME conductor. There is only one other motor connected.

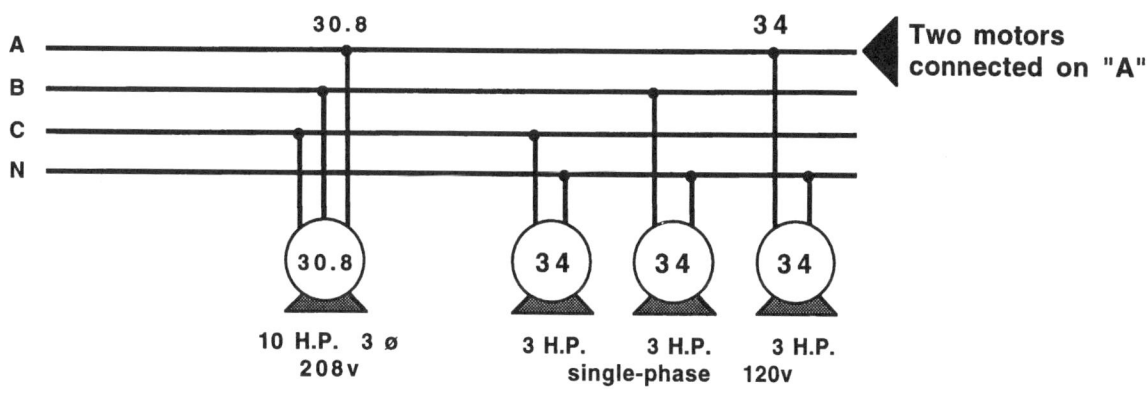

Question #40 Solution:

#600 kcmil x 2 = 1200 kcmil. Table 250.66 over 1100 kcmil = **#3/0 copper** to the water pipe.

250.66(A) = **#6 copper** to the ground rod.

KEY:

Table 250.66 for over 1100 kcmil a #3/0 would be the largest copper conductor required.

Section 250.66(A) states a #6 copper is the largest conductor required to a ground rod.

Question #41 Solution:

Table 314.16(B)

#14 = 2 cubic inches x 4 = 8 cu.in.
#12 = 2.25 cu.in. x 4 = 9 cu.in.
#10 = 2.5 cu.in. x 5 = 12.5 cu.in.
#8 = 3 cu.in. x 3 = 9 cu.in.
                    38.5 cubic inches required

Table 314.16(A) = **4 11/16" x 2 1/8" square box**.

KEY:

Table 314.16(A) lists the cubic inch capacity for boxes. A square box 4 11/16" x 2 1/8" has a cubic inch capacity of 42.

Question #42 Solution:

1500 sq.ft. x 3va	=	4500va
Small appl. 2 x 1500va	=	3000
Laundry	=	1500
		9000va

Table 220.11 demand:
1st 3000va @ 100%	=	3000va
Next 6000va @ 35%	=	2100
Water heater	=	6000
14kw range	=	8800
Dryer	=	5500
		25,400va

(b) 20,000 - 26,000va

KEY:

The washing machine has already been counted in the laundry @ 1500va.
• 1500va is the minimum, if the washing machine load had been larger than 1500va, then it would have replaced the laundry @ 1500va minimum.

Question #43 Solution:

250 receptacles at 180va (220.3B11) = 45,000va or 45 kva

Table 220.13 demand: 1st 10kva @ 100%   = 10 kva
                      Next 35kva @ 50%   = 17.5 kva
                                           27.5 kva

(a) 25 - 100 kva

KEY:

Table 220.13 demand applies to receptacles in NONdwelling units.

Question #44 Solution:

$$\frac{3{,}320{,}000 \text{ va}}{480v \times 1.732 \times 80\%} = 4991.8 \text{ amps}$$

(d) 5000 amps

KEY:

3320 kva = 3,320,000va. Since it's 3ø use 1.732 in the formula.

Question #45 Solution:

30a x 6 times (430.110C3) = **180 amps.**

KEY:

430-110(C3) states the locked-rotor current shall be *assumed* to be 6 times the full-load current of the motor.

Question #46 Solution:

50 amps x 200% (630.12A) = **100 amps.**

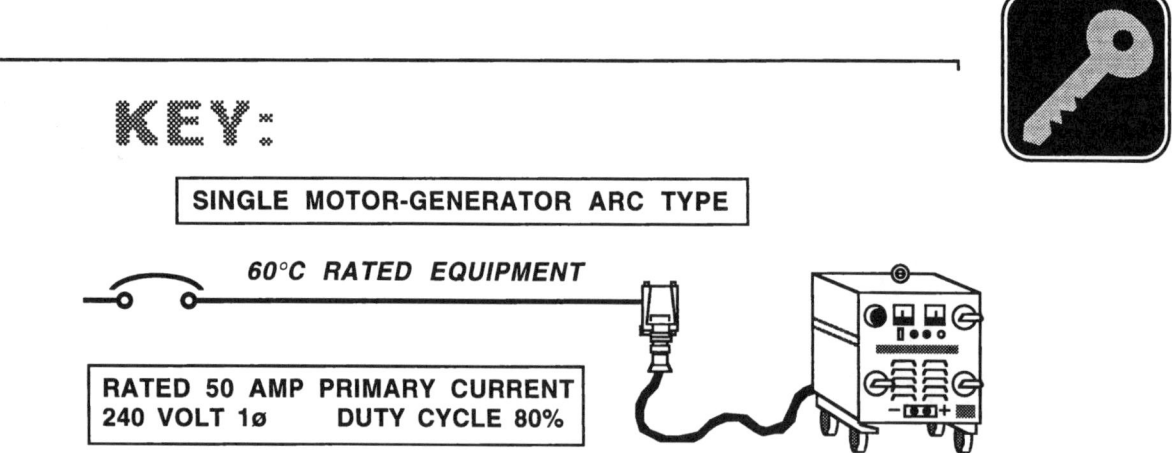

Article 630 Part C is for Motor-Generator Arc Welders
630.12(A) requires 200% for the overcurrent protection.

Question #47 Solution:

Table 5 area of square inches:

#750 kcmil THW	= 1.1652 sq.in. x 8 =	9.3216 sq.in.
#1000 kcmil THW	= 1.4784 sq.in. x 6 =	8.8704 sq.in.
		18.192 square inches

KEY:

Table 392.10(A) shows the allowable cable fill area for *single conductor* cables in ladder or ventilated trough cable trays 2000v or less.

Question #48 Solution:

#10 THWN-2 = 40 ampacity x 1.04 correction factor = **41.6 ampacity**

KEY:

Table 310.13$_5$ THWN listed wire types designated with suffix "-2" shall be permitted at 90°C wet or dry location.

A #10 THWN-2 in Table 310.16 would have a normal ampacity of 40 amps from the 90°C column.

The ambient temperature is only 70°F, the correction is 1.04 which is a 4% increase in the ampacity for the cooler ambient.

Question #49 Solution:

    50 units x 15' x 30' x 2va (T.220.3A)   =   <u>45,000va</u>

    T. 220.11 demand:
    1st 20,000va @ 50%   =   10,000va
    Next 25,000va @ 40%   =   10,000va
    Hallways 4000sq.ft. x .5va x *125%*   =   <u>2,500va</u>
                                                     22,500va

KEY:

The question is asking for the demand for a wing which is 50 units.

Table 220.3(A) lists the guest rooms at 2va and the hallways at 1/2 va. The guest rooms are subject to a demand from Table 220.11. The hallway load must be increased 25% for continuous loading per 215.2(A).

Guest rooms are recognized the same as a dwelling noncontinuous loading.

Question #50 Solution:

$E = I \times R = 8a \times 30\Omega = $ **240 volts**

KEY:

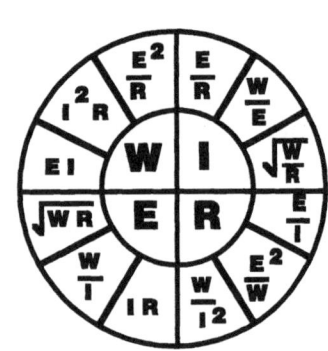

The total current is 18 amps in parallel with R1 showing 10 amps.
R2 would be 18 - 10 = 8 amps. R2 shows a resistance of 30 ohms. 8a x 30Ω = 240v.

ANSWERS - CALCULATIONS II

Question #51 Solution:

$$\frac{680,000va}{20,000 \text{ sq.ft.}} = 34va \text{ per sq.ft.}$$

Table 220.34:

1st 3va @ 100%	=	3va
Next 17va @ 75%	=	12.75va
Next 14va @ 25%	=	3.5va
		19.25va

19.25va x 20,000 sq.ft = 385,000va/1000 = **385 kva**

KEY:

The first step is to divide the total load by the total square footage and then apply the demand factor to this number.

Optional method for a school is Code section 220.34. Table 220.34 permits a demand to be applied above 3va.

Question #52 Solution:

(d) in series with the H.C. normally open contact

KEY:

When the start button is pressed the "M" coil will energize closing the "HC" interlock placing it in series with the stop button. When the stop button is pressed it will open the circuit to the "M" coil.

•Note: The "HC" contact should be designated "M" as it is physically attached to the "M" frame. This is an error made by the person writing this exam question. I show calculations exactly as asked and mention any errors I feel were made.

Question #53 Solution:

200 units x 5 kw = 100 kw x 25% Table 220.18 demand = **250 kw**

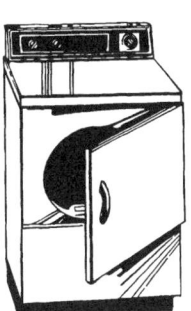

KEY: **5 KW MINIMUM**

220.18 requires a minimum of *5 kw* for a clothes dryer. Table 220.18 allows a demand for over four dryers.

Question #54 Solution:

1/3 + 1/4 + 1/8

8/24 + 6/24 + 3/24 = **17/24 horsepower**

KEY:

The question is checking to see if you know how to add fractions. The first step is to find the lowest common denominator which would be 24.

Question #55 Solution:

```
1500 kva
480/277v  3ø
3.5% Impedance
```

$$\frac{1500 \times 1000}{480v \times 1.732} = 1804 \text{ amps}$$

1804 amps x **100** = 180,400a/3.5 = 51,542 amps

(c) 51,000 - 75,000

KEY:

Formula for short circuit current:

$$I = \frac{\text{transformer full load amps} \times \mathbf{100}}{\text{transformer \% impedance}}$$

Example: 167 kva, 3ø, 208/120v, 2% impedance.

167 kva = 167,000va

$$\frac{167,000va}{208v \times 1.732} = 463.5 \text{ or } 464 \text{ amps}$$

464a x **100** = 46400/2 = 23,200 amperes

ANSWERS - CALCULATIONS II

Question #56 Solution:

The 6000va three-phase 208v motor divides 1/3 on each phase. The three 5000va heaters would divide 1/2 per phase since the single-phase voltage is 208v, the heaters would connect from L1-L2, L2-L3, and L3-L1. The three 150va lights are single-phase 120 volt, 120v connects from line to neutral. Since we have three 150va loads, we can equally balance them one per phase.

PHASE A	PHASE B	PHASE C
2000va	2000va	2000va
2500va	2500va	2500va
2500va	2500va	2500va
150va	150va	150va
7150va	7150va	7150va

L1 amps　　　　**L2 amps**　　　　**L3 amps**

$\dfrac{7150va}{120v} = 59.5$　　$\dfrac{7150va}{120v} = 59.5$　　$\dfrac{7150va}{120v} = 59.5$

KEY:

- Checkpoint: WYE-connected phase amps = line amps,

  Total va = 7150va x 3 phases = $\dfrac{21{,}450va}{208v \times 1.732}$ = 59.5 line amps.

Question #57 Solution:

24v/480v = .05 or **5%**

KEY:

Transformer impedance (Z) helps determine what the short circuit current will be at the transformer secondary. Transformer impedance is determined as follows:

The transformer secondary is shorted. Voltage is applied to the primary which causes full load current to flow in the secondary. The applied voltage divided by the rated primary voltage is the impedance of the transformer.

ANSWERS - CALCULATIONS II

Question #58 Solution:

Table 220.19 Column C:

35 ranges = 15kw + 35 kw (1kw for each range) = **50 kw**

KEY:

Table 220.19 Column C lists the maximum demand for household ranges not over 12 kw in size.

Question #59 Solution:

$$\frac{250 \text{ ohms}}{100 \text{ feet}} = 2.5 \text{ ohms per foot}$$

$$\frac{10 \text{ ohms total resistance}}{2.5 \text{ ohms per foot}} = \textbf{4 feet of wire}$$

KEY:

The first step is to solve the resistance per foot of wire. Then divide the total resistance by the resistance per foot will give you the length of wire.

Question #60 Solution:

$$I = E/R \quad 120v/15\Omega = \textbf{8 amps}$$

KEY:

When the switch is closed current follows the path of least resistance. The 10Ω, 15Ω, 20Ω, and 25Ω are bypassed as the switch has less resistance. Only the 5Ω and 10Ω are in the circuit. Since they are in series the resistances add for a total of 15Ω.

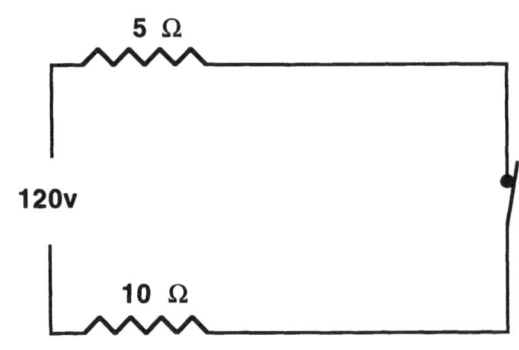
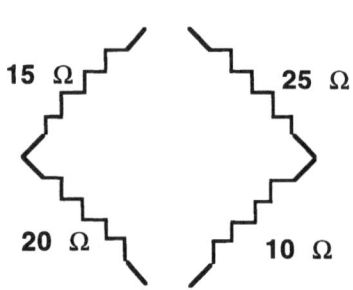

Question #61 Solution:

Optional method multifamily 220.32

900 sq.ft x 3va	=	2700va
Small appl. 2 x 1500va	=	3000va
Laundry	=	1500va
Heat	=	6000
Range	=	8000
Water heater	=	4500
Dishwasher	=	1200
		26,900va

26,900va x 24 units = 645,600/1000 = 645.6 kw

645.6 kw x 35% (24 units) demand Table 220.32 = 225.96 kw

(b) 200 - 400 kw

KEY:

220.32(C3) all appliances go at *nameplate* optional method. Table 220.32 permits a demand of 35% for 24 units.

ANSWERS - CALCULATIONS II

Question #62 Solution:

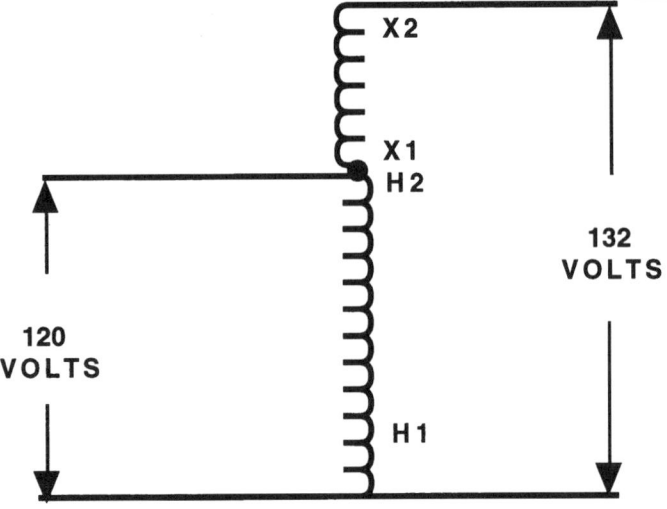

KEY:

When voltage is stepped **up**, part of the winding is the primary, and all of the winding is the secondary.

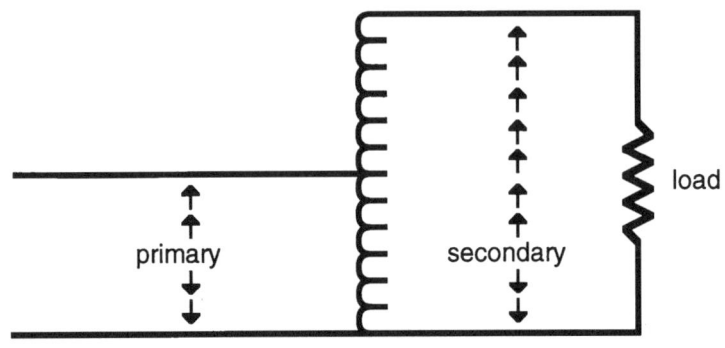

Question #63 Solution:

Code section 366.7(A) ampacity of aluminum is 700 amps per square inch.

.25" x 1.5" = .375 square inches x 700 amps = **262.5 amps.**

KEY:

Bus Bar Square inch area = Width x Thickness

Ampacity = 1000 amps per square inch for copper
700 amps per square inch for aluminum

Question #64 Solution:

12 units x 800 sq.ft. x 3va	=	28800va
Small appliance 24 x 1500va	=	36000va
Laundry 12 x 1500va	=	18000va
		82,800va

Table 220.11 Demand:
First 3000va @ 100%   =   3000va
Next 79,800va @ 35%   =   27930va
T.220.19 Column C = 27kw x 70% = 18,900
                                                        49,830va

49,830va/230v = 217 amps   220.22: First 200a @ 100% = 200a
                                                        Next   17a @ 70%   =   12a
                                                                                    212 amps on neutral

(a) 212 amps

KEY:

The question is on the *neutral* load. Section 220.22 permits a reduction on a neutral load in excess of 200 amps.

Question #65 Solution:

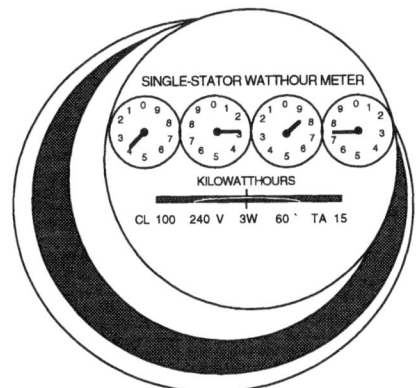

The meter is read at the beginning of the month and reads *3287*.

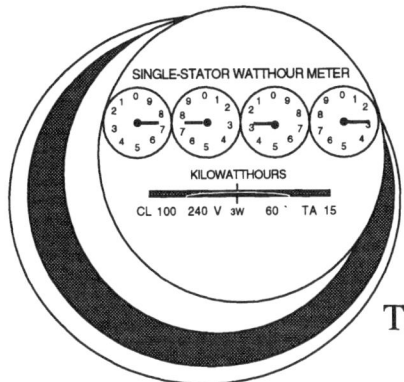

The meter is read at the end of the month and reads *7722*.

The KWH would be 7722 - 3287 = **4435.**

KEY:

The reading of the meter dials is from left to right.
The pointer must have passed the number to count it.

At 8 cents per KWH the bill would amount to 4435 KWH x .08 = $354.80.

Question #66 Solution:

Table 430.148 1ø F.L.C.:

5 hp 1ø, 230v = 28 amps

4 hp 1ø, 230v = ?

3 hp 1ø, 230v = 17 amps

The difference in amps between a 3 hp and a 5 hp is: 28a - 17a = 11 amps.

Divide 11 amps by 2 = 5.5 amps per hp.

3 hp = 17a + 5.5a = **22.5 amps** F.L.C. for a 4 hp motor.

KEY:

Code section 430.6(A) states: Where a motor is marked in amperes, but not horsepower, the horsepower rating shall be assumed to be that corresponding to the value given in Table 430.148, *interpolated* if necessary.

Question #67 Solution:

Cost = watts x hours used x rate per hour / 1000

W = E x I   120v x 2.5a = 300 watts

300w x 8 hours x .08 / 1000   =   .192 cents per day

.192 cents x 30 day month = **$5.76 cost** per month.

KEY:

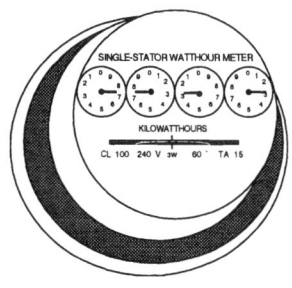

8 cents per kwh is pretty close to the average throughout the U.S.

At 8 cents per kwh a 100 watt light bulb can operate for 10 hours continuously for a cost of 8 cents.

ANSWERS - CALCULATIONS II

Question #68 Solution:

$E = I \times R$ 3 amps x 40 ohms = 120 volts

$W = I^2 R$

R1 = 3a x 3a x 40Ω = 360 watts

$I = E/R$ 120v/80Ω = 1.5 amps

R3 = 1.5a x 1.5a x 80Ω = 180 watts

R2 = 1.5a x 1.5a x 80Ω = 180 watts (R2 is also 80Ω to have a total R of 20)

(a) R1 would overheat at 360 watts

KEY:

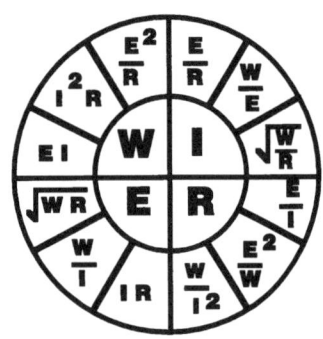

Rt = 20Ω $\dfrac{R2 \times R3}{R2 + R3}$ = $\dfrac{80 \times 80}{80 + 80}$ = $\dfrac{6400}{160}$ = 40

$\dfrac{R1 \times 40}{R1 + 40}$ = $\dfrac{40 \times 40}{40 + 40}$ = $\dfrac{1600}{80}$ = 20 R total

ANSWERS - CALCULATIONS II

Question #69 Solution:

Section 314.16(B):

3 - black wires
3 - white wires
1 - grounding wires
1 - clamps
2 - duplex receptacle
<u>2 -</u> single-pole switch
12 wires

12 x 2.25 cu.in. (Table 314.16B) = **27 cubic inch** required.

KEY: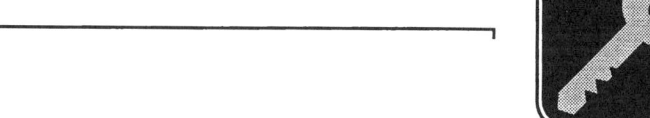

Code section 314.16(B1) states you count one for each wire.
314.16(B2) states you count one wire for clamps.
314.16(B4) states you count two for a device.
314.16(B5) states you count one for one *or more* grounding wires.

Question #70 Solution:

Secondary line amps = 80, same current in the secondary phase of a wye connected transformer 80 amps. The ratio from 480 phase to 120 phase is 4/1. The current in the primary phase would be 80/4 = 20 amps. Delta connected 20a x 1.732 = **34.64 amps** primary line.

KEY:

WYE CONNECTED

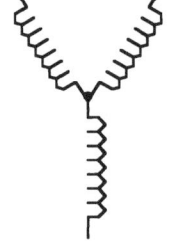

PHASE AMPS = LINE AMPS

PHASE VOLTS x 1.732 = LINE VOLTS

OR

$\dfrac{\text{LINE VOLTS}}{1.732}$ = PHASE VOLTS

DELTA CONNECTED

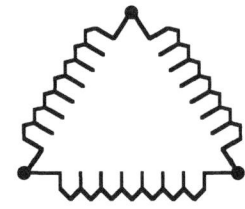

PHASE AMPS x 1.732 = LINE AMPS

OR

$\dfrac{\text{LINE AMPS}}{1.732}$ = PHASE AMPS

PHASE VOLTS = LINE VOLTS

Question #71 Solution:

Table 310.16:

#12 THHN = 30a x .82 (correction factor for 122°F) = 24.6a

24.6a x 80% (T.310-15(B2a) adjustment for 6 current-carrying wires) = **19.68 ampacity.**

KEY:

The 30 ampacity can be used for a 90°C conductor for derating purposes, but a 90°C conductor can only be loaded to the ampacity of a 60°C or 75°C conductor per section 110.14(C).

ANSWERS - CALCULATIONS II

Question #72 Solution:

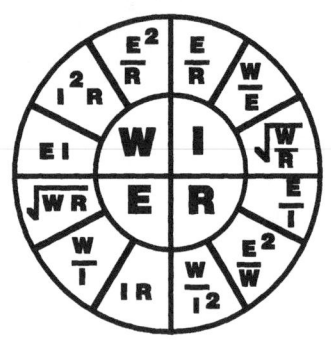

$W = I^2R$ = $I = E/R = 120v/10\Omega = 12$ phase amps

12a x 12a x 10Ω = 1440w per phase x 3 phases = **4320 watts.**

KEY:

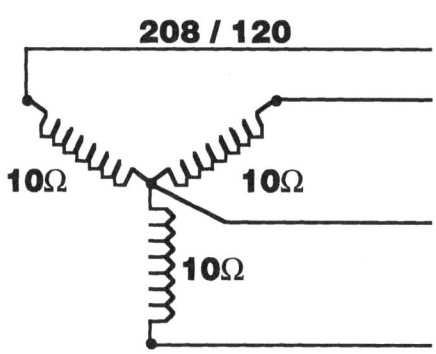

Wye connected phase amps equal line amps.

208v x 12a x 1.732 = 4323 watts.

Question #73 Solution:

For **10** ranges, find the best balance per phase:

A PHASE	B PHASE	C PHASE
4 ranges	3 ranges	3 ranges

Take the largest number of ranges on one phase (4 on A PHASE) times the number of connections (two). 4 ranges x 2 = **8 appliances**. This is the advantage of a 3-phase over a single-phase, instead of calculating 10 appliance loads, you now have **8** appliance loads.

Table 220.19 Column C for 8 ranges = **23 kw** maximum demand.

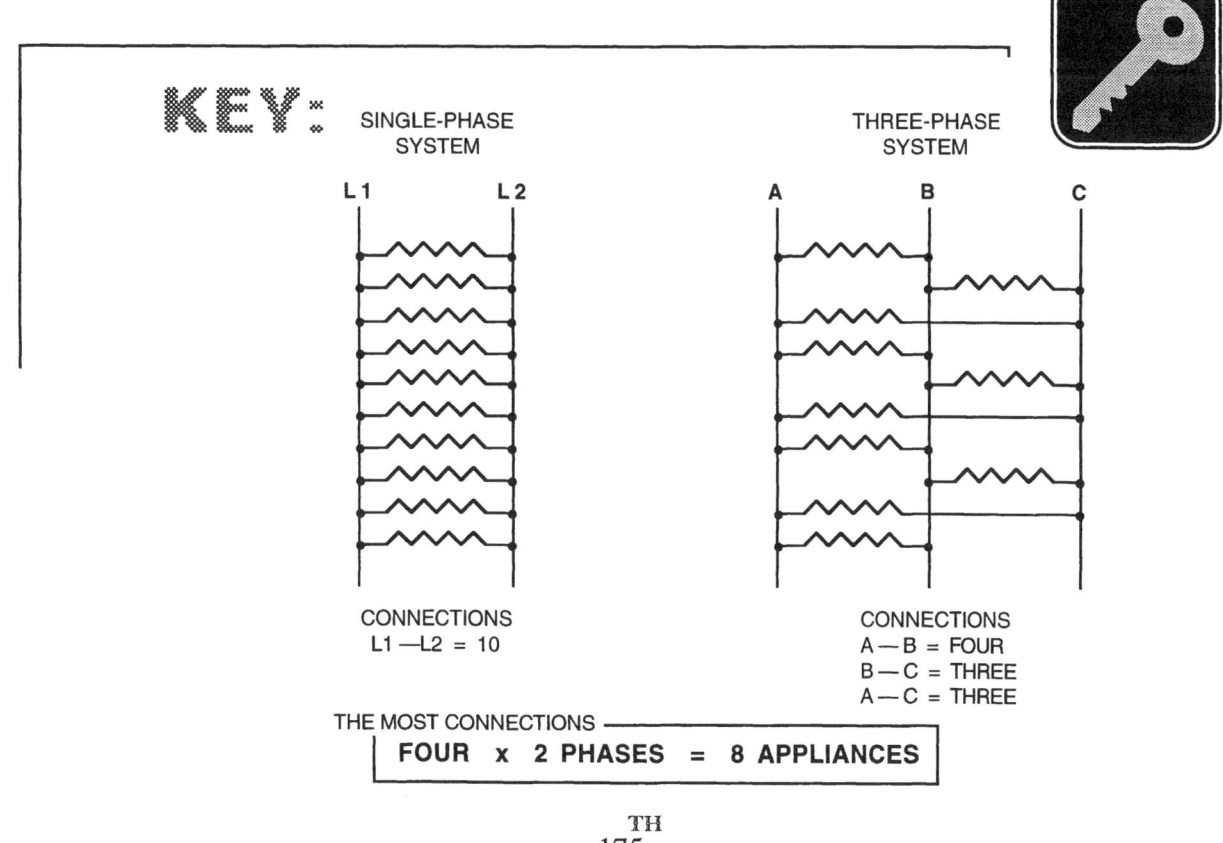

ANSWERS - CALCULATIONS II

Question #74 Solution:

Primary voltage is 240 volts with a 3/2 ratio.

240v/3 = 80 volts x 2 secondary ratio = **160 volts** secondary.

KEY:

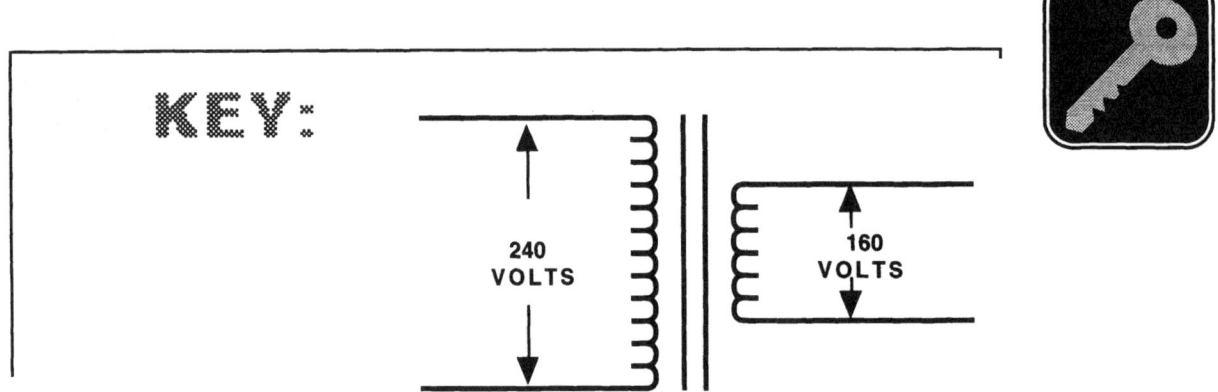

The primary winding has more turns than the secondary. Each turn represents voltage. A 3/2 ratio means the secondary has two-thirds the voltage or 66%. 160v/240v = 66%.

Question #75 Solution:

All of the following EXCEPT **(d) the "STP" coil de-energizing**

KEY:

Although motor control questions are not actually calculations, they are asked in the calculation part of the exam. The electrical exam applicant is expected to be able to read schematic diagrams and know the meaning of the symbol designations.

ANSWERS - CALCULATIONS II

Question #76 Solution:

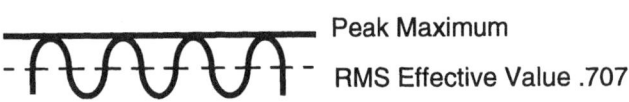

Peak Maximum
RMS Effective Value .707

METER READS
RMS EFFECTIVE VALUE

480v/.707 = **679 volts maximum**

KEY:

The AC circuit will have to be increased to 1.414 amperes before it will produce the **same heating effect** as will one ampere of DC current. Similarly, the peak voltage is 1.414 times the RMS voltage.

RMS VALUE = MAXIMUM x .707   MAXIMUM VALUE = RMS/.707

Question #77 Solution:

#8 THHN   50a x .71 (630.11A) = 35.5 required ampacity. Table 310.16 #10 THHN has an ampacity of 40, but the 60°C ampacity must be selected requiring a #8.

KEY:

Code section 110.14(C) requires equipment to be listed at 90°C to use conductors at 90°C ampacity. The equipment in this question is rated 60°C.

ANSWERS - CALCULATIONS II

Question #78 Solution:

$$\frac{50 \text{ amp load}}{.80 \text{ (T.310-15B2a)}} = 62.5 \text{ required ampacity}$$

Table 310.16 = **#6 THWN**

KEY:

The formula to find wire size: Load/correction factors

Section 310-15(B4c) requires counting the neutral conductor when it's supplying nonlinear loads such as fluorescent lights. Table 310-15(B2a) for four current-carrying conductors requires an ampacity adjustment of 80%.

Question #79 Solution:

(d)	A	B
	3000	1500
	1920	1920
	1920	1920
	2500	2500
	4000	4000
	1200	

KEY:

The 4 kw clothes dryer must be calculated at a 5 kw minimum per section 220.18. 2500 watts would be balanced between each line.

ANSWERS - CALCULATIONS II

Question #80 Solution:

1st step find PF = w/va   watts = 746 x 100 hp = 74,600 watts true power
va = E x I   2300v x 49a x 1.732 = 195,196va
PF = 74,600w/195,196va = *38% power factor*

2nd step the inductive vars in the circuit are calculated using this formula:

vars = $\sqrt{va^2 - w^2}$ = 195,196va x 195,196va = 38,101,478,416va
74,600w x 74,600w = 5,565,160,000w now subtract the watts from the volt amps = 38,101,478,416va - 5,565,160,000w = 32,536,318,416 now press the square root button on your calculator $\sqrt{}$ = 180,378 vars

3rd step find the required va to produce a 95% power factor. va required = w/pf = 74,600w/.95 = 78,526va required

4th step to find the required inductive vars to produce the va required:

vars = $\sqrt{va^2 - w^2}$ = 78,526va x 78,526va = 6,166,332,676va
74,600w x 74,600w = 5,565,160,000w   now subtract the watts from the volt amps = 6,166,332,676va - 5,565,160,000w = 601,172,676 now press the square root button on your calculator $\sqrt{}$ = 24,518.8 or 24,519 vars

KEY:

5th step the present inductive vars are 180,378; to find the needed vars to produce a reactive power of 24,519 vars subtract the amount of needed vars from the present amount of vars = 180,378 vars - 24,519 vars = *155,859 vars or 155.8 or* **156 kvars** *needed to raise to 95% power factor.*

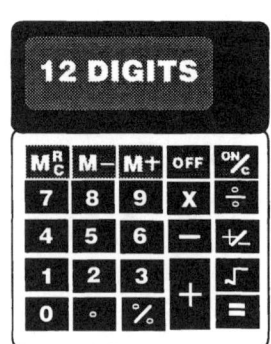

•Note: I used a **12** digit calculator to solve this example. A standard calculator that has 8 digits won't work.

TH

Question #81 Solution:

Section 314.28(A2): 6 x 3" = 18" + 2" = **20"**

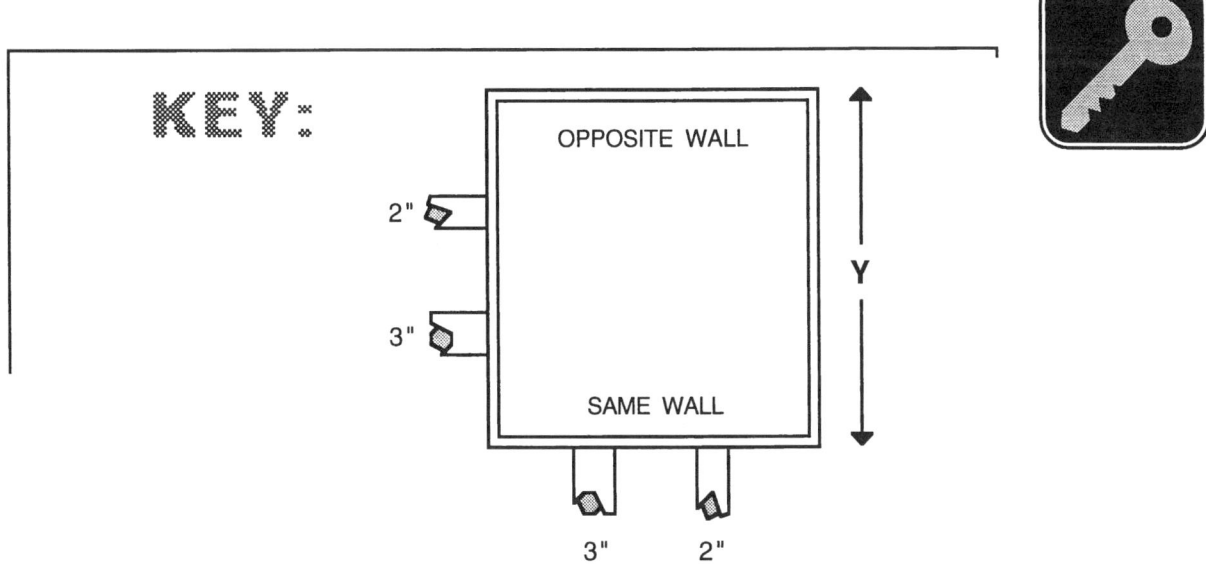

Largest raceway on **same wall** (3" x 6 = 18") **plus** all other raceways on the **same wall** (2").
3" x 6 = 18" + 2" = 20".   Dimension **"Y" = 20" minimum**.

ANSWERS - CALCULATIONS II

Question #82 Solution:

(c) 8

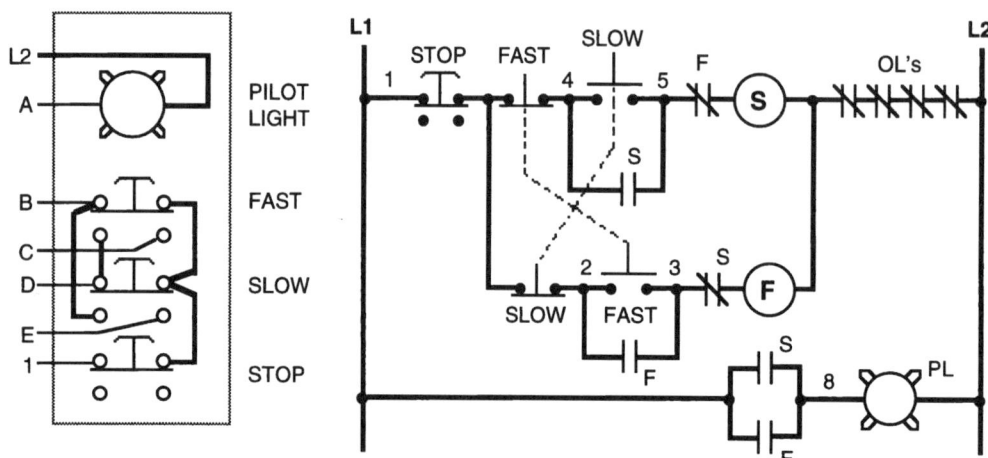

KEY:

Conductor "A" connects to one side of the pilot light. The other side of the pilot light is connected to "L2".

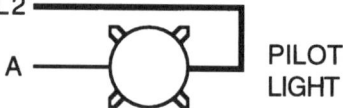

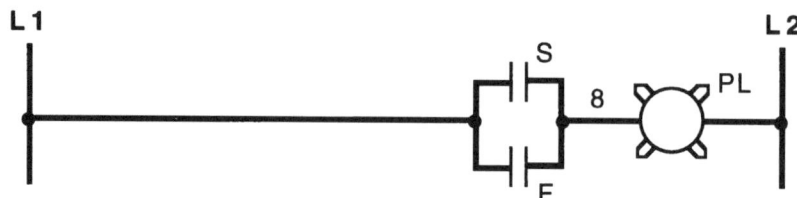

Question #83 Solution:

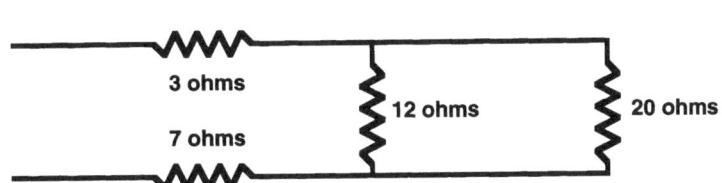

The 12Ω and 20Ω are in parallel: $\dfrac{12 \times 20}{12 + 20} = \dfrac{240}{32} = 7.5\Omega$

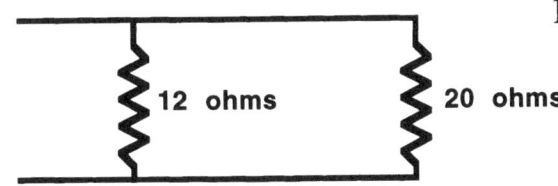

Now the circuit looks like this:

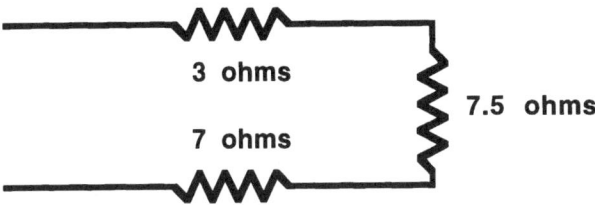

In series the resistance adds: $3\Omega + 7.5\Omega + 7\Omega +$ **17.5** R total.

KEY:

Always start at the end of the circuit and reduce it to its simplest form.

Question #84 Solution:

Article 555.12:

```
15 x 15 amps = 225 amps
12 x 30 amps = 360 amps
 8 x 50 amps = 400 amps
35             985 amps
```

Table 555.12: For 35 receptacles 60% demand

985 amps x 60% = **591 amps minimum required**.

KEY:

Article 555 is for Marinas and Boatyards. The demand factors of Table 555.12 may be inadequate in areas of extreme hot or cold temperature with loaded circuits for heating, air-conditioning, or refrigeration equipment.

• Due to the long lengths of circuits in most marinas you should always calculate for voltage drop.

Question #85 Solution:

630.31(B) FPN 3   300 x 20 = 6000/216,000 = .0277777 x 100 = 2.77 or **2.8%**.

KEY:

The duty cycle is the percentage of the time during which the welder is loaded.

Question #86 Solution:

$$\frac{7200v}{240v} = 30 \text{ ratio}$$

$$\frac{1800 \text{ turns}}{30 \text{ ratio}} = \textbf{60 turns} \text{ in secondary winding.}$$

KEY:

7200 volt primary has 1800 turns would be 7200/1800 = 4 volts per turn. The secondary voltage is 240. Divide 240/4v per turn = 60 turns.

ANSWERS - CALCULATIONS II

Question #87 Solution:

2000 sq.ft. x 3va (Table 220.3A)	=	6000va
Small appliance 2 x 1500va	=	3000
Laundry	=	1500
		10,500va

Table 220.11 demand:
1st 3000va @ 100%	=	3000va
Next 7500va @ 35%	=	2625
		5625va

5625va/240v = **23.4 amps**.

KEY:

Code section 240.23 states: Where a change occurs in the size of the ungrounded (hot) conductor, a similar change shall be permitted to be made in the size of the grounded (neutral) conductor.

Question #88 Solution:

First step is to find the circuit resistance:

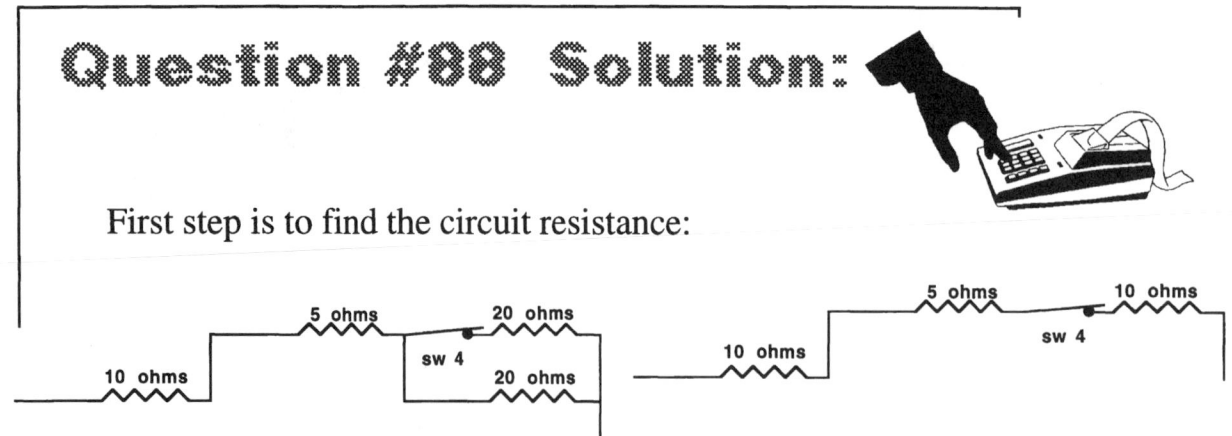

The two 20Ω resistors in parallel would reduce to a 10Ω. Total R = 25Ω.

Next step is to find I = E/R = 225v/25Ω = 9 amps current flow.

VD = I x R = 9 amps x 10Ω = **90 volts** dropped across a 10Ω resistor.

KEY:

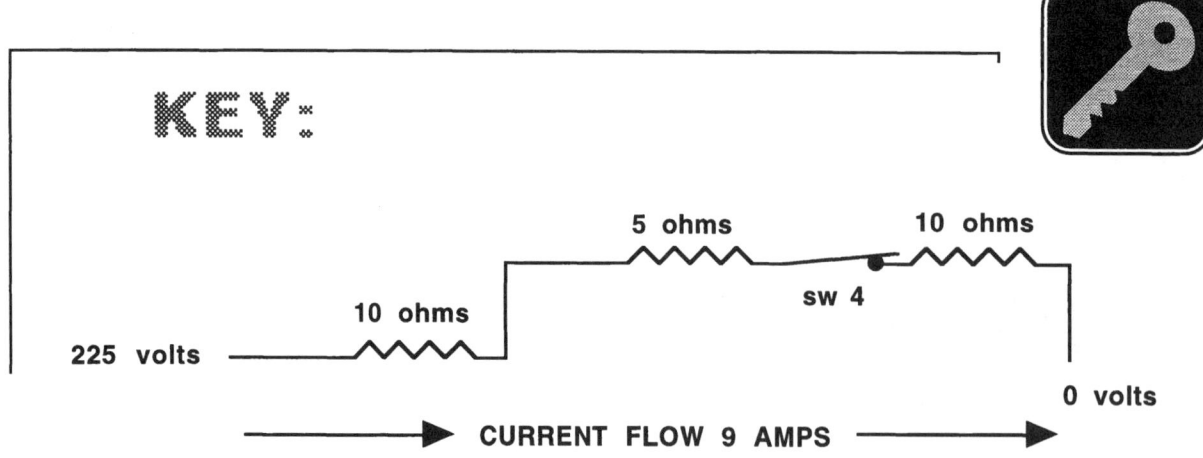

One must always understand to push a current through a resistance it takes a pressure (voltage). The entire 225 volts drop to push the 9 amps through these resistances. The voltage is replaced every 1/60th of a second. The drop at a 10Ω was 10Ω x 9a = 90vd, the drop across the 5Ω is 5Ω x 9a = 45vd, the drop across the other 10Ω is 10Ω x 9a = 90vd. 90vd + 45vd + 90vd = the applied voltage 225v.

Question #89 Solution:

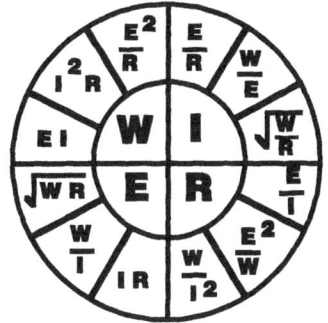

I = E/R = 120v/10Ω = 12 phase amps

VA = E x I = 120v x 12a = 1440va per phase

1440va x 3 phases = 4320va/1000 = **4.32 kva**.

KEY:

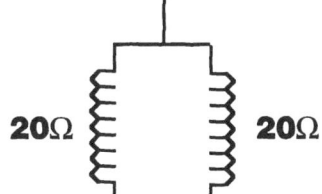

A phase has two equal 20Ω loads in parallel.

To find the phase resistance: 20Ω/2 resistances = 10Ω total phase resistance.

Total resistance in parallel for *equal* resistances = $\dfrac{\text{RESISTANCE of ONE}}{\text{NUMBER of RESISTANCES}}$

ANSWERS - CALCULATIONS II

Question #90 Solution:

- 1 - 25 kw fryer
- 1 - 2 kw toaster
- 1 - 6 kw booster heater
- 1 - 24 kw oven
- 1 - 10 kw dishwasher
- 1 - 2 kw bun warmer

Table 220.20:
69 kw x 65% = 44.85 kw but the two largest loads are 25kw + 24kw = 49kw

lights at 125% and are rated 4 kva = 4,000 x 125% (215.2A1) = 5000va

62 general use receptacles at 180va = 11,160va Table 220.13 =
1st 10,000va @ 100% = 10,000va
Next 1160va @ 50% =   580va
                      10,580va

49,000 + 5,000 + 10,580 = 64,580/208v x 1.732 = **179 amps.**

KEY:

Code section 220.20 states: However, in no case shall the feeder demand be less than the sum of the largest two kitchen equipment loads.

Table 220.13 permits a demand for nondwelling unit receptacle loads over 10 kva or 10,000va.

Question #91 Solution:

60 x 1200va = 72,000va x 75% (220.17) = 54,000va

54,000va/240v = **225 amps**.

KEY:

Section 220.17 permits a demand factor of 75% for *four or more* fastened in place appliances in dwelling units.

ANSWERS - CALCULATIONS II

Question #92 Solution:

Solution: $I_n = \sqrt{I^2A + I^2B + I^2C - (IA\ IB) - (IB\ IC) - (IC\ IA)}$

The formula at first looks very difficult, but really it's not.

Everything under the square root sign must be done first, which means:

current in A squared = 20 x 20 = 400
+ current in B squared = 35 x 35 = 1225
+ current in C squared = 40 x 40 = 1600
                                    3225 total (call this total "X")

Now the right side of the formula shows:

current in A x current in B = 20 x 35 = 700
current in B x current in C = 35 x 40 = 1400
current in C x current in A = 40 x 20 = 800
                                         2900 (call this total "Y")

Now subtract total "Y" from total "X" = 3225 total X
                                      — 2900 total Y
                                         325

Now extract the square root by pressing the $\sqrt{\ }$ button on your calculator. The answer **18.02** amps is the unbalanced current flowing in the neutral. 18.027756 is the square root of 325.

KEY:

The formula looks difficult, but it is actually a very easy calculation using your calculator.

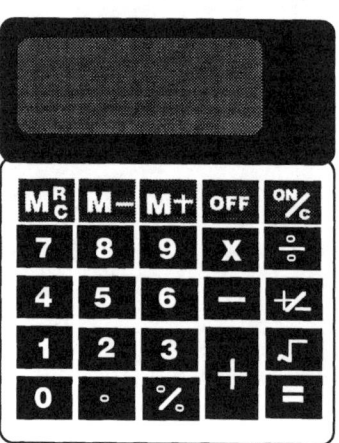

Question #93 Solution:

Table 220.19 Note 4: = **8 kw.**

KEY:

Note 4 states: For *one* wall-mounted oven OR *one* counter-mounted cooktop the unit shall be *nameplate*.

Table 220.19 demand can be used for a branch circuit to one *range*, but not a oven or cooktop.

•*A range is a combination of both an oven and a cooktop.*

Question #94 Solution:

$$VD = \frac{2 \times K \times D \times I}{CM} \quad \text{or} \quad VD = I \times R$$

$$VD = \frac{2 \times 12.6029 \times 150' \times 9.8a}{6530 \text{ cm}} = \textbf{5.6742 volts}$$

or

$$VD = 9.8a \times .579\Omega = 5.6742 \text{ volts}$$

KEY:

To find K factor = $\frac{R \times CM}{1000'}$ = $\frac{1.93\Omega \text{ (Table 8)} \times 6530\text{cm (Table 8)}}{1000'}$ = 12.6029

To find R = Table 8 #12 solid = 1.93Ω per k/ft

1.93Ω x .300' of wire = .579Ω

Question #95 Solution:

30 feet show window x 200va (220.12A) x 125% (210-19A) = 7500va

7500va/120vv = 62.5 amps.

KEY:

Chapter 9 example #D3 shows the store show window at continuous load.

ANSWERS - CALCULATIONS II

Question #96 Solution:

The question is on the *primary* which is delta.

$$\frac{50,000 \text{ va}}{480\text{v} \times 1.732} = \textbf{60 line amps}$$

or you can calculate the delta phase amps and convert them to line amps:

50,000va/ 3 phases = 16,666va per phase   phase voltage = 480v

16,666va/480v = 34.7 delta phase amps

34.7a x 1.732 = 60 delta line amps.

KEY:

DELTA CONNECTED

PHASE AMPS x 1.732 = LINE AMPS

OR

$$\frac{\text{LINE AMPS}}{1.732} = \text{PHASE AMPS}$$

PHASE VOLTS = LINE VOLTS

ANSWERS - CALCULATIONS II

Question #97 Solution:

240 sq.ft. x 2va (Table 220.3A) x 100 units	=	48,000va
Table 220.11 demand:		
1st 20,000va @ 50%	=	10,000va
Next 28,000va @ 40%	=	11,200va
Hallways 2000 sq.ft. x .5va x 125% (215.2A1)	=	1,250va
Beauty shop 400 sq.ft. x 3va x 125%	=	1,500va
Office 1200 sq.ft. x 3.5va x 125%	=	5,250va
Office 1200 sq.ft. x 1va (unknown recpts.)	=	1,200va
		30,400va

KEY:

The guest rooms are not considered continuous load. A demand factor applies to the guest room load only.

The office 3.5va is for lighting which would be a continuous load and require a 25% increase per section 215.2(A1).

The office 1va for unknown receptacle loads is not a continuous load, as receptacles in general are not loaded for 3 hours continuously.

Question #98 Solution:

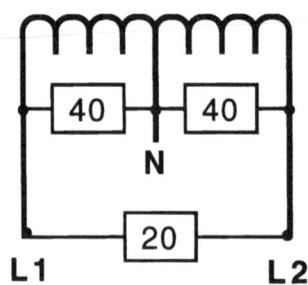

40,000w/120v = **333 amps.**

KEY:

The question is on the *neutral*. The total load is 100kw, but 20kw is 240 volts and does *not* connect to the neutral. Subtract 20kw from the 100kw and you have 80kw left that is balanced (80/2=40) on the neutral. The worst condition would be if you lost either line. The neutral would be sized to 333 amps which is the maximum load it would ever carry.

ANSWERS - CALCULATIONS II

Question #99 Solution:

120v x 20a = $\dfrac{\text{2400va per circuit}}{\text{180va per outlet}}$ = **13 receptacles**

KEY:

Code section 220.3(B11) requires 180va per outlet. A 20 amp circuit has a capacity of 2400va.

• There is no limit on the number of receptacles on a branch circuit in a *dwelling*. Table 220.3(A) *No other load is required. The 3va per square foot covers all receptacle loads in a dwelling.

Question #100 Solution:

Table 430.148: 13.2a F.L.C. x 175% (Table 430.152) = 23.1 amps

240.6 next higher standard size is 25 amps, but it is not a choice.

The answer is **20 amps**.

KEY:

The Code permits in 430.52(C1) ex.1 that you can go up to the **next** higher standard size which would be 25 amps. But, 25 amps is NOT one of the answer choices and the Code does NOT permit to go up to a 30 amp as that would be two sizes higher. You are allowed to go up ONLY one size higher. You would drop down to a 20 amp, which is NOT a violation of the Code.

Question #101 Solution:

20,000 sq.ft. x 3va x 125%	75000
4,000 sq.ft. x 1va x 125%	5000
2,000 sq.ft. x 2va x 125%	5000
10 kw outside lights 10,000 x 125%	12500
5 kw stage lighting (assume noncontinuous)	5000
200 receptacles 200 x 180va = 36000	
Table 220.13: First 10 kva @ 100% =	10000
Remaining 26 kva @ 50% =	13000
2 - 14 kw ranges = 28 kw x 65% (T.220.20)	18200
2 - 6 kw ovens = 12 kw x 65% (T.220.20)	7800
3 - 4 kw fryers = 12 kw x 65% (T.220.20)	7800
1 - 12 kw water heater = 12 kw x 65% (T.220.20)	7800
1 - 3 kw dishwasher = 3 kw x 65% (T.220.20)	1950
1 - 6 kw heater = 6 kw x 65% (T.220.20)	3900
2 - 2 kw toasters = 4 kw x 65% (T.220.20)	2600
2 - 1/2 hp fans 9.8a x 120v = 1176va x 2 = 2352va x 65%	1529
2 - 3/4 hp fans 13.8a x 120v = 1656va x 2 = 3312 x 65%	2153
4 - 10 hp A/C units 55a x 208v = 11440va x 4	45760
40 kw heat (omit)	
Largest motor 11440va x 25%	2860
	227,852va

KEY:

Apply an increase of 25% to continuous loads per 215.2(A1) 180va for receptacles with a demand from Table 220.13 for receptacle loads over 10 kva. Apply Table 220.20 demand to all kitchen equipment that pertains to the operation of cooking. Omit the smallest load (heat) per 220.21. Largest motor in hp is to be increased 25% per 430.24.

227,852va/108v = 1095 amps. Use 1200 amp single-phase service.

Tom Henry's
THE ELECTRICIANS BOOKSTORE

READ THE BOOKS THE ELECTRICIAN READS....

THE LIBRARY
SAVE $204

MASTERCOMBO #4

BEST VALUE!

ALL **27** of *Tom Henry's* electrical books for **$15.89** per book when purchased as a library!

VIDEOS
THE BLOCKBUSTER
By Tom Henry

 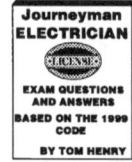  

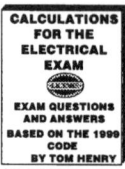   

ITEM #513 - JOURNEYMAN SERIES includes tapes #501 through #509. A total of 9 videos for $299.00

ITEM #514 - MASTER SERIES includes all 12 videos for $399.00

*VIDEOS CAN BE ORDERED SEPARATELY USE ITEM # BELOW $39.95 each

EACH VHS TAPE IS 75-120 MINUTES

THE EXAM — Item #501
OHMS LAW - THEORY — Item #502
VOLTAGE DROP & RESISTANCE — Item #503
AMPACITY CORRECTION FACTORS — Item #504

  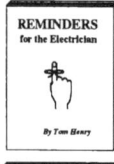 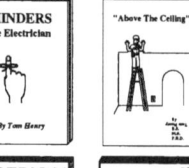

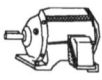  

MOTORS — Item #505
COOKING EQUIPMENT DEMAND FACTORS — Item #506
DWELLING-RESIDENTIAL SERVICE SIZING — Item #507

   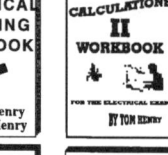

BOX and CONDUIT SIZING — Item #508
SINGLE-PHASE TRANSFORMERS — Item #509
THREE-PHASE TRANSFORMERS — Item #510

COMMERCIAL-MULTI-FAMILY SERVICE SIZING — Item #511
MOTOR CONTROL - SWITCH CONNECTIONS — Item #512

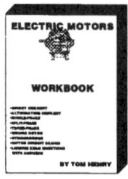  

ITEM #224 - Master Combo #4 - **Start your own library!** 27 of Tom Henry's study-aid books and instructor guides. Every school program and electrical contractor should have this combo. The complete study guide. Why take the exam more than once? Properly prepare yourself with this combo. Save $204 total price is **$429.00**

☎ **CALL TODAY!** ☎
1-800-642-2633
E-Mail: tomhenry@code-electrical.com
http://www.code-electrical.com